Springer
Berlin
Heidelberg
New York
Barcelona
Budapest
Hong Kong
London
Milan
Paris
Tokyo

Thomas F. Lüscher (Ed.)

The Endothelium in Cardiovascular Disease

Pathophysiology, Clinical Presentation and Pharmacotherapy

With 75 Figures

Springer

Professor Dr. med. Thomas F. Lüscher
Abteilung Kardiologie, Inselspital
3010 Bern
Switzerland

Library of Congress Cataloging-in-Publication Data

The endothelium in cardiovascular disease : pathophysiology, clinical presentation, and pharmacology / T.F. Luscher (ed.).
p. cm.
Includes bibliographical references and index.
ISBN 3-540-59352-7 (alk. paper)
1. Vascular endothelium--Pathophysiology. 2. Endothelins--Pathophysiology. 3. Nitric oxide--Pathophysiology. I. Lüscher, Thomas F. (Thomas Felix)
[DNLM: 1. Endothelium, Vascular-- physiopathology.
2. Endothelins--physiology. 3. Endothelins-- antagonists & inhibitors. 4. Endothelium-Derived Relaxing Factor.
5. Cardiovascular Diseases--physiopathology. WG 500 E565 1995]
RC669.9.E53 1995
616.1--dc20
DNLM/DLC
for Library of Congress 95-31212
CIP

ISBN 3-540-59352-7 Springer-Verlag Berlin Heidelberg New York

Book production: PRO EDIT GmbH, D-69126 Heidelberg
Cover design: Springer-Verlag, Design & Production
Typesetting and printing: Zechnersche Buchdruckerei, D-67346 Speyer
Binding: Schäffer, D-67269 Grünstadt

SPIN: 10500840 23/3132- 5 4 3 2 1 0 – Printed an acid-free paper

Preface

Already described by Malpighi several hundred years ago, the endothelium has long been considered to be like inert wallpaper covering the inner lumen of all blood vessels. Its role in the regulation of permeability in the microcirculation was recognized later, but only recently has the important regulatory role of endothelial cells in the circulation been appreciated. In the 1970s Moncada and Vane observed that endothelial cells are a source of prostacyclin, a substance which inhibits platelet aggregation and causes vascular relaxation. Full recognition of the endothelium as an important cardiovascular regulator, however, did not come until Robert Furchgott's seminal paper in 1980 describing endothelium-derived relaxing factor produced within the cells. The nature of this endothelium-derived relaxing factor was not certain for several years. In 1987 Moncada and coworkers demonstrated that nitric oxide may account for the biological activity of the endothelium-derived relaxing factor described by Robert Furchgott. Since then the amount of literature on the endothelium as a regulator of vascular tone and growth as well as platelet function and coagulation has exploded. It is now clear that endothelial cells not only produce nitric oxide and prostacyclin, but most likely also an as yet unidentified hyperpolarizing factor. Furthermore, endothelial cells have been shown to produce potent constrictor factors such as endothelin-1, a 21-amino acid peptide, cyclooxygenase products such as prostaglandin H2 and thromboxane A2, and possibly also components of the renin-angiotensin system. The capacity for producing relaxing and contracting factors establishes the endothelium as an important regulator of vascular tone. Most interestingly, several of these substances also affect circulating blood cells such as platelets and monocytes. Indeed, platelets release substances which can stimulate the endothelium to produce nitric oxide and prostacyclin. Nitric oxide and prostacyclin, on the other hand, are able to deactivate platelets and prevent aggregation and adhesion on the cells. Hence, nitric oxide and prostacyclin may play an important protective role in the vessel wall. More recently it has been suggested that nitric oxide also acts as an inhibitor of proliferation and migration of vascular smooth muscle cells.

Due to its strategic anatomical position between the circulating blood and vascular smooth muscle, the endothelium is a primary target and mediator of cardiovascular disease. Indeed, endothelial dysfunction has been described in a variety of disease states such as hypercholesteremia, hypertension, diabetes, and atherosclerosis. As certain of these dysfunctions of the endothelium occur rather early in the disease process, pharmacological modulation of its function may be a promising approach to prevent the development of various forms of cardiovascular disease.

In this monography a number of well-known clinicians and scientists in the field of endothelial cell research have contributed chapters covering different aspects of endothelial cell physiology, dysfunction, and perspectives of pharmacotherapy. It is the aim of the editor to stimulate further research in this area and to offer the notion that endothelial mediators and their modulation are now ready to be used not only in basic cardiovascular research but also in clinical research and practice.

Bern, May 8, 1995 THOMAS F. LÜSCHER

Contents

List of Contributors

DR. J. C. BURNETT
Professor of Medicine and Physiology
Cardiovascular Diseases and Internal Medicine, Mayo Clinic,
200 First Street Southwest, Rochester, MI 55905, USA

DR. A. L. CLAVELL
Cardiovascular Diseases and Internal Medicine, Mayo Clinic,
200 First Street Southwest, Rochester, MI 55905, USA

DR. H. L. ELLIOTT
University of Glasgow, Department of Medicine and Therapeutics,
Gardiner Institute – Western Infirmary, Glasgow G11 6NT, U.K.

DR. L. FERNANDEZ
Cardiovascular Preventive Center, Department of Internal Medicine,
Hospital Universitario "San Carlos",
Complutense University of Madrid, 28040 Madrid, Spain

DR. A. FERNANDEZ-CRUZ
Professor of Medicine
Cardiovascular Preventive Center, Department of Internal Medicine,
Hospital Universitario "San Carlos",
Complutense University of Madrid, 28040 Madrid, Spain

DR. R. FERNANDEZ-DURANGO
Cardiovascular Preventive Center, Department of Internal Medicine,
Hospital Universitario "San Carlos",
Complutense University of Madrid, 28040 Madrid, Spain

DR. C. J. FERRO
Clinical Research Centre, Department of Medicine,
University of Edinburgh, Western General Hospital,
Edinburgh EH4 2XU, U.K.

DR. G. GAZZANO
Centro di Fisiologia Clinica e Ipertensione, Ospedale Maggiore,
Via F. Sforza, 35, 20122 Milano, Italy

Dr. F. Ghio
Centro di Fisiologia Clinica e Ipertensione, Ospedale Maggiore,
Via F. Sforza, 35, 20122 Milano, Italy

Dr. H. Haller
Professor of Medicine
Franz-Volhard-Klinik, Universitätsklinikum Rudolf Virchow,
Wiltbergstraße 50, 13125 Berlin, Germany

Dr. W. G. Haynes
Clinical Research Centre, Department of Medicine,
University of Edinburgh, Western General Hospital,
Edinburgh EH4 2XU, U.K.

Dr. M. Hogie
Service de Pharmacologie, Hopital de Bois Guillaume,
CHU de Rouen, 76031 Rouen Cedex, France

Dr. J. Ibarra
Cardiovascular Preventive Center, Department of Internal Medicine,
Hospital Universitario "San Carlos",
Complutense University of Madrid, 28040 Madrid, Spain

Dr. D. Langleben
3755 chemin Cote-Ste-Catherine,
Cardiology Division, Jewish General Hospital,
Montreal, PQ H3T 1E2, Canada

Dr. T. F. Lüscher
Professor of Medicine
Cardiology, University Hospital, Inselspital,
3010 Bern, Switzerland

Dr. A. Morganti
Professor of Medicine
Centro di Fisiologia Clinica e Ipertensione, Ospedale Maggiore,
Via F. Sforza, 35, 20122 Milano, Italy

Dr. P. Morrison
U-Gene Research, Bolognalaan 40, 3508 CJ Utrecht, The Netherlands

Dr. J. Moya
Cardiovascular Preventive Center, Department of Internal Medicine,
Hospital Universitario "San Carlos",
Complutense University of Madrid, 28040 Madrid, Spain

Dr. G. Noll
Cardiology, University Hospital, Inselspital, 3010 Bern, Switzerland

Dr. R. Patino
Cardiovascular Preventive Center, Department of Internal Medicine,
Hospital Universitario "San Carlos",
Complutense University of Madrid, 28040 Madrid, Spain

Dr. T. J. Rabelink
Department of Nephrology and Hypertension (F03.226), University Hospital Utrecht, P.O. Box 85500, 3508 GA Utrecht, The Netherlands

Dr. V. Richard
Service de Pharmacologie, Hopital de Bois Guillaume, CHU de Rouen, 76031 Rouen Cedex, France

Dr. M. Roldan
Department of Ophthalmology, Hospital Universitario "San Carlos", Complutense University of Madrid, 28040 Madrid, Spain

Dr. A. Salvetti
Professor of Medicine
I. Clinica Medica, University of Pisa, Via Roma, 67, 56100 Pisa, Italy

Dr. U. Scherrer
Assistant Professor of Medicine
Department of Internal Medicine, BH 10.642, CHUV, 1011 Lausanne, Switzerland

Dr. E. L. Schiffrin
Professor of Medicine
Clinical Research Institute of Montreal, 110, Pine Avenue West, Montreal, Quebec H2W 1R7, Canada

Dr. D. J. Stewart
Head, Cardiology Division, Room 712B,
St. Michael Hospital,
30, Bond Street, Toronto, Ontario M5B 1W8, Canada

Dr. S. Taddei
I. Clinica Medica, University of Pisa, Via Roma, 67, 56100 Pisa, Italy

Dr. C. Thuillez
Professor of Medicine
Service de Pharmacologie, Hopital de Bois Guillaume, CHU de Rouen, 76031 Rouen Cedex, France

Dr. H. Vierhapper
Associate Professor of Medicine
Clinical Division of Endocrinology and Metabolism,
University Clinic for Internal Medicine III,
Währinger Gürtel 18–20, 1090 Wien, Austria

Dr. D. J. Webb
Senior Lecturer
Clinical Research Center, Department of Medicine, University of Edinburgh, Western General Hospital, Edinburgh EH4 2XU, U.K.

Dr. R. R. Wenzel
Cardiology, University Hospital, Inselspital, 3010 Bern, Switzerland

The Endothelium as a Regulator of Vascular Tone and Growth

T. F. Lüscher and G. Noll

Endothelium-Derived Relaxing Factors

L-Arginine Nitric Oxide Pathway

The endothelium, when stimulated by neurotransmitters, hormones, substances derived from platelets and the coagulation system, can evoke relaxations (Fig. 1; [1, 2]). The mediator of these responses is a diffusible substance with a half-life of a few seconds, the so-called endothelium-derived relaxing factor (EDRF; [2]) which has recently been identified as the free radical nitric oxide (NO; [3]). NO is formed for L-arginine by oxidation of its guanidine-nitrogen terminal [4]. The catalyzing enzyme NO synthase is constitutively expressed and exists in several isoforms in endothelial cells, platelets, macrophages, vascular smooth muscle cells and in the brain [5]. In endothelial cells, NO synthase gene expression – although constitutively activated –

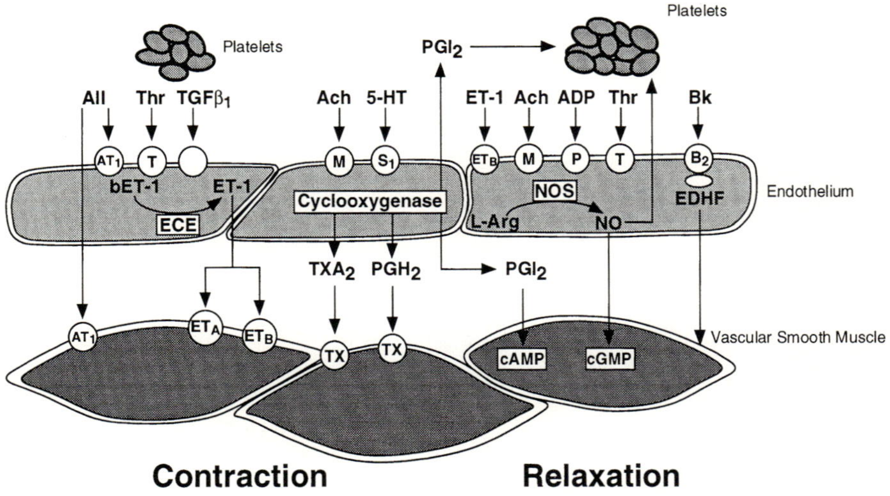

Fig. 1. Endothelium-derived vasoactive substances. The endothelium is a source of relaxing (*right*) and contracting factors (*left*). *AA*, arachidonic acid; *AT*, angiotensin; *Ach*, acetylcholine; *ADP*, adenosine diphosphate; *Bk*, bradykinin; *cAMP/cGMP*, cyclic adenosine/guanosine monophosphate; *EDHF*, endothelium-derived hyperpolarizing factor; *5-HT*, 5-hydroxytryptamine (serotonin); *ET*, endothelin-1; *L-Arg*, L-arginine; *NO*, nitric oxide; O_2^-, superoxide radical; PGI_2, prostacyclin; $TGF\beta_1$, transforming growth factor β_1; *Thr*, thrombin; TXA_2, thromboxane A_2; *Circles* represent receptors

is upregulated by shear stress and estrogens (Fig. 2; B. S. Oemar, personal communication). The activity of the enzyme is inhibited by the circulating amino acid asymmetrical dimethyl-arginine (ADMA) which accumulates in patients with renal failure [6].

In addition to receptor-operated mechanisms mechanical factors (i.e., shear forces exerted by the circulating blood) induce endothelium-dependent vasodilation, an important adaptive response of the vasculature during exercise. Indeed, while normally reactive hyperemia is associated with flow-dependent vasodilation of large conduit arteries also in humans, infusion of the inhibitor of NO formation L-N^G-monomethyl arginine (L-NMMA) unmasks a vasoconstriction (most likely due to the unoppossed effects of the sympathetic nervous system; Fig. 3; [7]). This response is immediate and therefore not regulated at the messenger RNA level, but rather involves stimulation of enzyme activity possibly via ion channels.

Particularly in the arterial, but not in the venous circulation, basal NO production can be noted. In isolated arteries, L-NMMA or other NO inhibitors induce endothelium-dependent contractions [8–10]. In perfused organs such as the isolated rabbit heart [11] or the perfused porcine eye (Fig. 4; [12]), NO inhibitors cause marked decreases in organ perfusion as they do in the human forearm circulation [13]. Adding NO inhibitors to the drinking water is followed by a pronounced and sustained hypertension in the rat [14–17]. Hence, due to the continuous production of NO the circulation is in a constant state of vasodilation. The intracellular mechanism by which NO causes relaxation in vascular smooth muscle cells involves formation of cyclic 3′,5′-guanosine monophosphate via the enzyme soluble guanylyl cyclase (Fig. 1; [18]). The fact that NO deficient animals have a markedly reduced

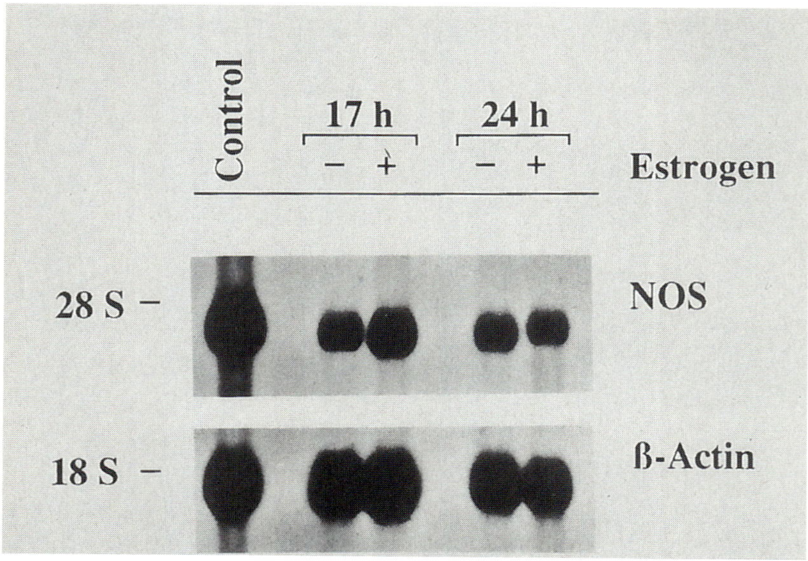

Fig. 2. Northern blot of constitutive nitric oxide synthase (cNOS) messenger RNA under control conditions and after pretreatment with 17β estradiol (unpublished observation). Note the transient increase in expression of cNOS at 17 h by 17β estradiol (17 h+)

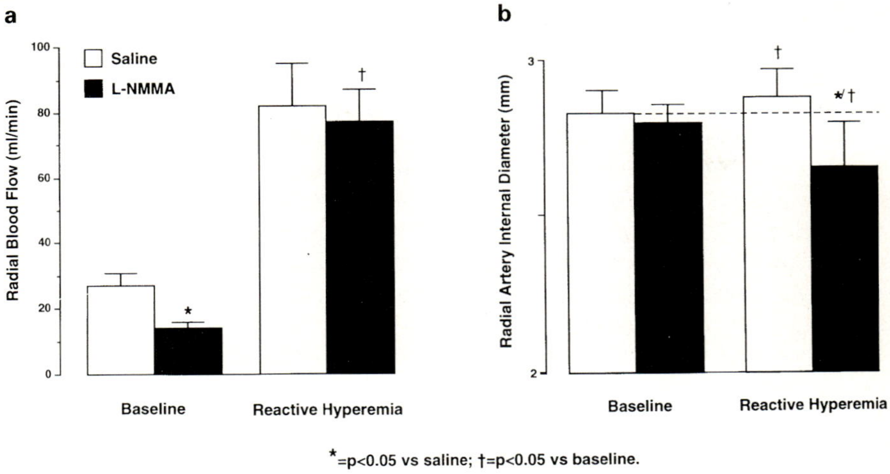

Fig. 3 a, b. Flow-dependent vasodilation in the human forearm circulation as measured by Doppler (radial blood flow; **a**) or B-mode ultrasound of the radial artery (internal diameter; **b**). Values are shown before and after infusion of L-N^G-monomethyl arginine (L-NMMA). Reactive hyperemia has been induced by occluding blood flow around the wrist and releasing thereafter. Note the increase in radial artery internal diameter in response to increased flow and the paradoxical vasoconstriction after inhibition of nitric oxide synthesis by L-NMMA. (Modified from [7])

survival due to target organ damage in the central nervous system, the heart and the kidney further suggests that NO exerts an important protective role in the circulation [14–17].

NO is released abluminally as well as luminally where it interacts with circulating blood cells and proteins (Fig. 1). Certain plasma proteins such as albumin become

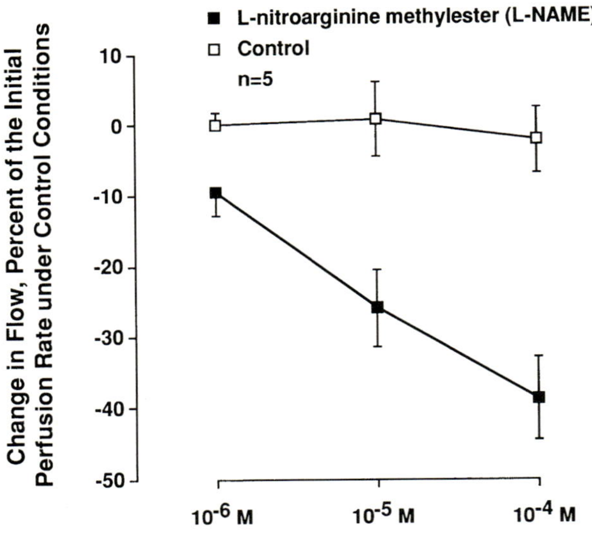

Fig. 4. Role of nitric oxide in the perfused porcine eye. Inhibition of nitric oxide formation by L-arginine methylester (L-NAME) leads to a concentration dependent decrease in ophthalmic flow. (Modified from [12])

nitrosylated and may act as circulating reservoirs of nitric oxide [19]. In platelets, an increase of intracellular 3′,5′-guanosine monophosphate is associated with a reduced adhesion and aggregation [20–22]. Platelets themselves posses a L-arginine/NO pathway which regulates their aggregability [22]. Most interestingly, platelets release substances such as adenosine diphosphate and triposphate as well as serotonin which activate the release of NO and prostacyclin from the endothelium (Fig. 1; [23, 24]). Furthermore, thrombin, the major enzyme of the coagulation cascade, stimulates the formation of NO by the endothelium [25]. Hence, at sites where platelets and the coagulation cascade are activated, intact endothelial cells immediately release NO and, in turn, cause vasodilation and platelet inhibition, thereby preventing vasoconstriction and thrombus formation (Fig. 1).

In addition, an inducible enzyme exists in vascular smooth muscle, endothelium and macrophages [26]. The enzyme is calcium-independent and produces large amounts of NO; it is induced by cytokines such as endotoxin, interleukin-1β and tumor necrosis factor and, hence, is activated in inflammatory processes and endotoxin shock [27]. In inflammatory tissue induction of NO production may contribute to the increased local perfusion (i.e., rubor) and act as an toxic molecule aiming at destroying microorganisms.

Prostacyclin

In addition to NO, prostacyclin is released by endothelial cells in response to shear stress, hypoxia and several substances (see above) which also release NO. Prostacyclin increases cyclic 3′,5′-adenosine monophosphate (cAMP) in smooth muscle and platelets [28]. Its platelet inhibitory effects are probably more important than its contribution to endothelium-dependent relaxation. Indeed, at least in the porcine coronary circulation prostacyclin contributes very little if at all to the regulation of vascular tone [29]. In platelets NO and prostacyclin synergistically inhibit platelet aggregation suggesting that the activity of both mediators is required to exert full antiplatelet activity [20].

Endothelium-Derived Hyperpolarizing Factor

In the coronary circulation, not all endothelium-dependent relaxations are prevented by inhibitors of the L-arginine pathway [29]. These NO-independent relaxations are even more prominent in intramyocardial vessels [10]. As under these conditions, vascular smooth muscle cells become hyperpolarized, an endothelium-dependent hyperpolarizing factor of unknown chemical structure has been proposed (Fig. 1; [30]). Indirect evidence suggests that it may be a product of the cytochrome P450 pathway, but C-type natriuretic peptide (CNP) may also be a candidate.

Endothelium-Derived Contracting Factors

Soon after the discovery of endothelium-derived relaxing factors, it became clear that endothelial cells also can mediate contraction, at least under certain conditions (Fig. 1; [1, 31]). Endothelium-derived contracting factors include the 21 amino acid peptide endothelin, vasoconstrictor prostanoids such as thromboxane A_2 and prostaglandin H_2 as well as components of the renin angiotensin system.

Endothelin-1

Endothelin exists in three isoforms, i.e., endothelin-1, endothelin-2, and endothelin-3. Endothelial cells produce exclusively endothelin-1 [32]. Translation of messenger RNA generates preproendothelin, which is converted to big endothelin; its conversion to the mature peptide endothelin-1 by the endothelin converting enzymes (ECE) is necessary for the development of full vascular activity (Fig. 1). Two isoforms of ECE have recently been cloned [33, 34]. The expression of messenger RNA and the release of the peptide is stimulated by thrombin, transforming growth factor β_1, interleukin-1, epinephrine, angiotensin II, arginine vasopressin, calcium ionophore and phorbol ester (Fig. 1; [32, 35]). In addition, hypoxia particularly at high altitude is associated with an increase in plasma endothelin levels which parallel the increase in systolic pulmonary pressure (Fig. 5; [36]). Treatment of healthy mountaineers with the calcium antagonist nifedipine does not affect plasma endothelin levels, but reduces systolic pulmonary pressure (Fig. 5; [36]).

Endothelin-1 causes vasodilation at lower and marked and sustained contractions at higher concentrations (Fig. 6; [32, 37]) which in the heart eventually lead to ischemia, arrhythmias and death. Intramyocardial vessels are more sensitive to the vaso-

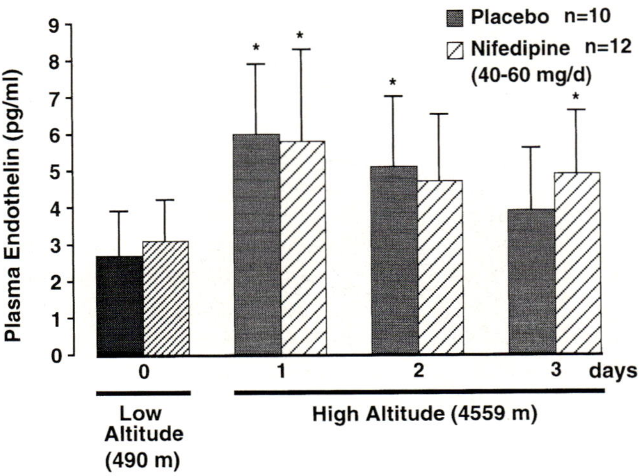

Fig. 5. Plasma endothelin levels in healthy mounteneers at low altitude and after exposure to high altitude. Values under placebo and nifedipine did not differ significantly. (Modified from [36])

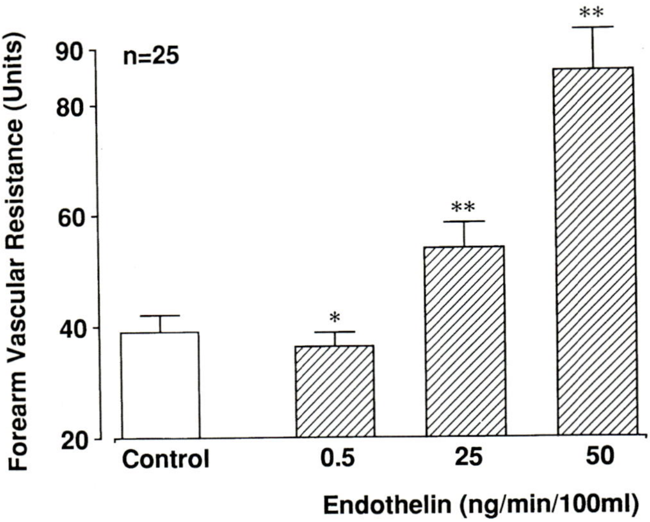

Fig. 6. Forearm vascular resistance in healthy human subjects before and after infusion of endothelin into the brachial artery. (* $p < 0.05$, ** $p < 0.01$). (Modified from [37])

constrictor effects of endothelin-1 than epicardial coronary arteries [38], suggesting that the peptide is particularly important in the regulation of flow. In general, veins are more sensitive to endothelin-1 than arteries [39].

The circulating levels of endothelin-1 are very low, suggesting that little of the peptide is formed physiologically (for review see [40]). This may be due to the absence of stimuli for endothelin production, the presence of potent inhibitory mechanisms or because it is released preferentially abluminally towards smooth muscle cells [41]. Three inhibitory mechanisms regulating endothelin production have been delineated: (1) cGMP-dependent inhibition [35]; (2) cAMP-dependent inhibition [42] and (3) an inhibitory factor produced by vascular smooth muscle cells [43]. After inhibition of the endothelial L-arginine pathway, the thrombin- or angiotensin-induced endothelin production is augmented [35]; on the other hand, nitrates and atrial natriuretic peptide (which activates particulate guanylyl cyclase) prevent thrombin-induced endothelin release via a cyclic GMP-dependent mechanism [35]. Endothelin can also release nitric oxide and prostacyclin from endothelial cells which as a negative feedback mechanism reduce endothelin production in the endothelium and its vasoconstrictor action in smooth muscle.

Two distinct endothelin receptors exist, the ET_A- and ET_B-receptor (Fig. 1; [44, 45]). Both are G protein-coupled receptors with seven transmembrane domains and are linked to phospholipase C and protein kinase C. Endothelial cells express ET_B-receptors linked to the formation of NO and prostacyclin which explains the transient vasodilator effects of endothelin when infused in intact organs or organisms. In vascular smooth muscle ET_A- and in part ET_B-receptors are mediating contraction and proliferation. Several endothelin receptor antagonists have been developed which are currently evaluated in normal subjects and patients [46].

Cyclooxygenase-Derived Contracting Factors

Particularly in veins, but also in the cerebral and ophthalmic circulation agonists such as arachidonic acid, acetylcholine, histamine and serotonin can evoke endothelium-dependent contractions which are mediated by thromboxane A_2 or prostaglandin H_2 (Fig. 1; [1]). Thromboxane A_2 and prostaglandin H_2 activate the thromboxane receptor in vascular smooth muscle and platelets and hence counteract the effects of NO and prostacyclin in both cells. In addition, the cyclooxygenase pathway is a source of superoxide anions which inactivate NO. Besides this indirect effect superoxide anions can also directly cause vasoconstriction.

Angiotensin Converting Enzyme

Finally, the endothelium regulates the activity of the renin angiotensin system; the angiotensin converting enzyme (ACE) which activates angiotensin I into angiotensin II is expressed on the endothelial cell membrane (Fig. 7; [47]). Angiotensin converting enzyme is identical to kinase II which breaksdown bradykinin. Whether or not other components of the renin angiotensin system are produced in endothelial cells is controversial. Angiotensin II can activate endothelial angiotensin receptors;

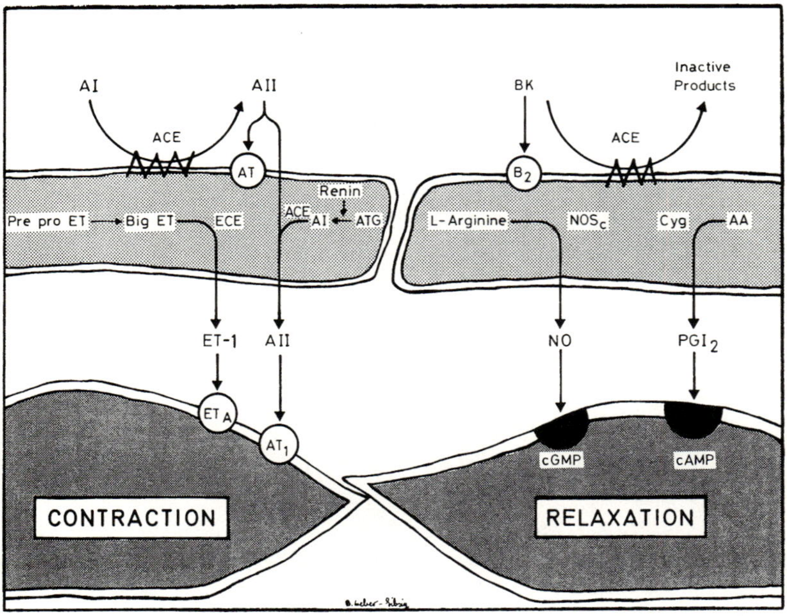

Fig. 7. Local vascular effects of angiotensin converting enzyme (*ACE*) inhibitors: ACE not only converts angiotensin I (*A I*) into angiotensin II (*A II*), but also inactivates bradykinin (*Bk*), a potent activator of nitric oxide (*NO*) and prostacyclin (*PGI_2*) production. *ET*, endothelin, *ECE*, endothelin converting enzyme, *NOS_c* constitutive nitric oxide synthase, *AA*, arachidonic acid, *Cyg*, cyclooxygenase; *circles* represent receptors

these receptors stimulate endothelin production and possibly that of other mediators such as plasminogen activator inhibitor.

Endothelium-Dependent Control of Vascular Structure

Removal of the endothelium, most commonly by a balloon catheter, causes immediate deposition of platelets and white blood cells. After days to weeks intimal hyperplasia develops at the site of injury [48]. This suggests that the endothelium also regulates vascular structure and that its presence is crucial to prevent migration and proliferation of vascular smooth muscle cells. Endothelial dysfunction on the other hand could be an important factor in atherosclerosis, restenosis and hypertensive vascular disease. Vascular structure is mainly determined by vascular smooth muscle cells and, in disease states, by white blood cells invading the intima.

Endothelial cells can have indirect and direct effects on vascular structure. NO and prostacyclin inhibit the adhesion of platelets to the vessel wall [49]. If at sites of endothelial dysfunction or denudation platelets do adhere to the blood vessel wall, they cause contraction (through the release of thromboxane A_2 and serotonin [23, 24]; and stimulate proliferation and migration of vascular smooth muscle cells (via the release of platelet-derived growth factor; Fig. 8; [50]). In addition, NO inhibits the adhesion of monocytes which are an important component of the atherosclerotic plaque and also capable to release growth factors and cytokines.

Furthermore, endothelial cells produce and release growth factors and inhibitors. Under physiological conditions it appears that growth inhibitors prevail and that this may explain why the blood vessel wall normally remains in a quiescent state

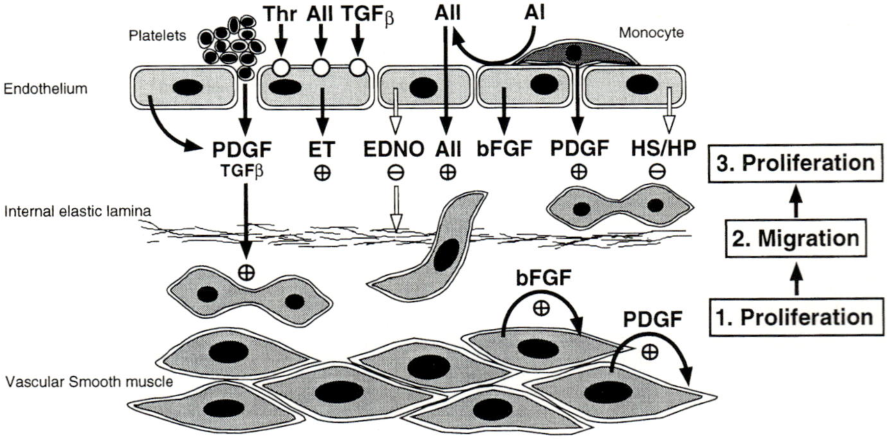

Fig. 8. Endothelium-dependent control of vascular structure. Normally the endothelium exerts a net inhibitory effect on mirgration and proliferation of vascular smooth muscle cells (*open arrows; –*). With endothelium dysfunction, platelets and monocytes adhere to the vessel wall and release growth factors, as does the endothelium (*black arrows; +*). *A II*, angiotensin II; *ET*, endothelin; *PDGF*, platelet-derived growth factor; *bFGF*, basic fibroblast growth factor; *TGFβ₁*, transforming growth factor beta₁; *EDNO*, nitric oxide; *HP/HS*, heparan sulfates; *Thr*, thrombin

and does not exhibit proliferative responses (Fig. 8). Heparane sulfates, NO and transforming growth factor β_1 are potent inhibitors of vascular smooth muscle migration and proliferation [51–53]. On the other hand, at least under certain conditions endothelial cells can produce various growth factors, in particular platelet-derived growth factor and epidermal growth factor (Fig. 8; [54, 55]). The fact that the endothelium can actively promote migration and growth of vascular smooth muscle cells through the release of specific molecules may explain why in spite of an intact endothelium in atherosclerosis proliferative responses of vascular smooth muscle cells ensue.

Endothelium in Cardiovascular Disease

The endothelium is that structure of the blood vessel wall which is most exposed to mechanical forces exerted by the blood and hormones and noxious substances therein. Morphologically, endothelial cell morphology changes with aging and disease; in particular an increased endothelial cell turnover and density, a marked heterogeneity in endothelial cell size and bulging of the cells into the lumen occurs. Endothelial cell denudation, however, is unusual under these conditions except in very late stages of atherosclerosis and at sites of rupture of an atherosclerotic plaque. Almost invariably associated with these changes in endothelial cell morphology are functional alterations and intimal thickening with accumulation of white blood cells, vascular smooth muscle cells and fibroblasts and matrix deposition.

Aging

The incidence and prevalence of all forms of cardiovascular disease increase with age. As these changes occur even in the absence of known cardiovascular risk factors, aging per se seems to alter vascular function. In most studies, endothelium-dependent relaxations decreased with aging (Fig. 9; [56]). In humans, the increase in coronary flow induced by acetylcholine infusion decreases with age [57]. Whether these changes are related to a dysfunction of muscarinic receptors and their signal transduction pathways or a decreased activity of NO synthase is still uncertain.

Endothelin production was found to increase with age in certain, but not all studies [58]. The response to endothelin, however, decreases with age [59] presumably due to receptor downregulation. Very little is know as regards the antithrombotic and antiproliferative properties of the endothelium with aging.

Hypertension

Endothelial dysfunction in hypertension could contribute to the increase in peripheral vascular resistance (particularly if it occurs in resistance arteries) and to the vascular complications of the disease (if present in large and medium-sized conduit arteries).

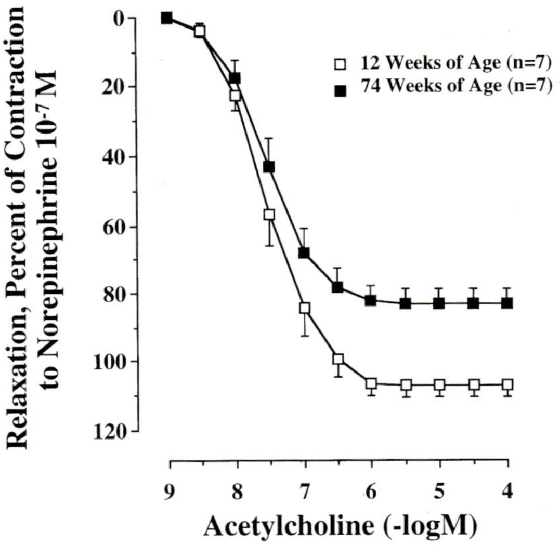

Fig. 9. Effects of aging on endothelium-dependent relaxations to acetylcholine in the aorta of Wistar Kyoto rats. (Modified from [56])

In most models of hypertension, high blood pressure is associated with reduced endothelium-dependent relaxations [1]. This defect is more dominant in certain than other blood vessels [60] and appears to occur as blood pressure rises and hence is a consequence rather than a cause of hypertension [61]. In hypertensive subjects, acetylcholine causes paradoxical vasoconstriction of epicardial coronary arteries. The increase in blood flow to acetylcholine in the forearm and coronary circulation was found to be decreased in all, but one study [62–64].

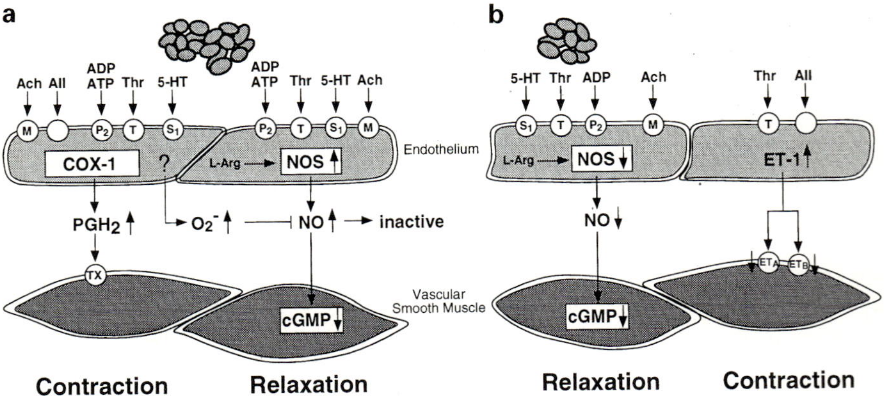

Fig. 10 a, b. Heterogeneity of endothelium dysfunction in hypertension. In spontaneous hypertension (**a**) nitric oxide synthase (*NOS*) activity is increased, but the biological activity of nitric oxide (*NO*) is reduced, possibly due to inactivation by superoxide radicals (O_2^-). In addition, the production of thromboxane A_2 (*TXA_2*) and prostaglandin H_2 (*PGH_2*) is increased. In contrast, in salt-induced hypertension (**b**), NO production is reduced and the endothelial production of vasoconstrictor prostaglandins is not present. *cGMP*, cyclic guanosine monophosphate. See legends to Figs. 1, 8 for other abbreviations.

The mechanism of endothelial dysfunction differs in different models of hypertension. In the spontaneously hypertensive rat, the activity of NO synthase is markedly increased, but inefficacious probably due to an increased deactivation of nitric oxide (Fig. 10; [65]). In addition, the endothelium of spontaneously hypertensive and ren-2 transgenic rats produces increased amounts of prostaglandin H_2 which offsets the effects of nitric oxide in vascular smooth muscle and platelets (Fig. 10; [1, 56]). Whether or not this occurs in man is uncertain; however, in the forearm circulation of patients with essential hypertension infusion of a cyclooxygenase inhibitor such as indomethacin enhances the vasodilation to acetylcholine [66]. In salt-induced hypertension of the rat, a reduced NO production is likely to occur (Fig. 10; [67]).

Endothelin plasma levels are normal in most patients with hypertension except in the presence of renal failure and atherosclerosis. Increased local vascular production of endothelin, however, remains a possibility as most of the peptide is released abluminally [41] and hence plasma levels do not necessarily reflect local tissue levels of endothelin. In the spontaneously hypertensive rat vascular endothelin production is reduced, but increased in DOCA-salt hypertensive rats[68].

Hyperlipidemia and Atherosclerosis

Endothelium-dependent relaxations are reduced in hyperlipidemia and atherosclerosis [69]. It is likely that low-density lipoproteins are a major determinant of this phenomenon. Indeed, incubation of isolated coronary arteries with oxidized but not native low-density lipoproteins selectively inhibits endothelium-dependent relaxations to serotonin, aggregating platelets and thrombin, while the response to bradykinin is unaffected [70]. A similar reduction of the response can be achieved by pertussis toxin or an inhibitor of NO formation suggesting that activation of the L-arginine pathway by G_i protein-coupled receptors becomes defective (Fig. 11; [70, 71]). As exogenous L-arginine improves or restores the reduced endothelium-dependent relaxations in the presence of oxidized low-density lipoproteins, it is possible that a reduced intracellular availability of L-arginine also contributes. The active component of low-density lipoproteins appears to be lysolecithine which mimics most of the effects of low-density lipoproteins. In line with the in vitro experiments, in the coronary artery of pigs with hypercholesteremia, a selective dysfunction of endothelium-dependent relaxations to serotonin, aggregating platelets and thrombin also is observed. In more advanced stages of atherosclerosis, a more generalized endothelial dysfunction occurs. Experiments in the hypercholesteremic rabbit aorta suggest that the overall production of NO is not reduced, but markedly augmented; however, NO is inactivated by superoxide radicals produced within the endothelium (Fig. 11; [72]). Similar observations have been made in rabbits with fully developed atherosclerosis. Under both conditions, the biologically active NO is markedly reduced, a fact also supported by bioassay experiments with coronary arteries of hypercholesteremic pigs [73].

Endothelin may be important in atherosclerotic vascular disease. In hyperlipidemia and atherosclerosis vascular endothelin production is increased (Fig. 4; [58]), while the expression of endothelin receptors is downregulated [74]. A most likely

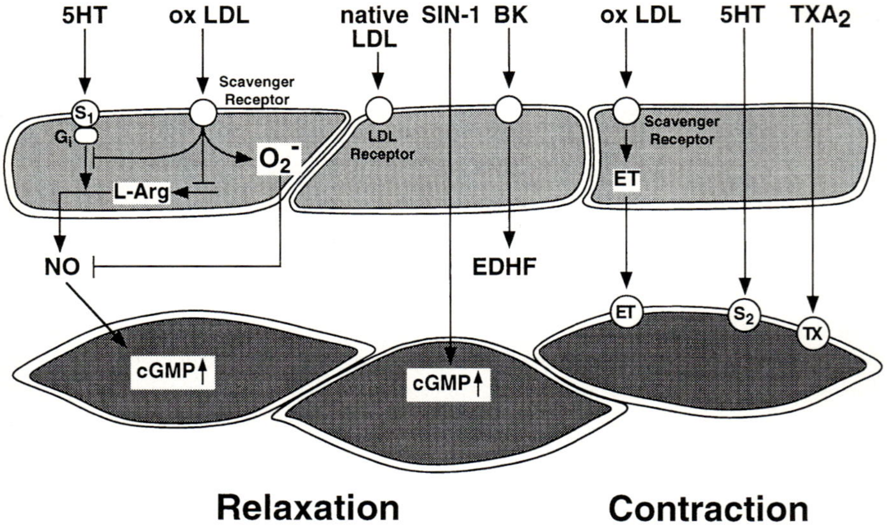

Fig. 11. Endothelium dysfunction in hyperlipidemia and atherosclerosis. The major component is oxidized low-density lipoprotein (*oxLDL*) which, by activating scavenger receptors, impairs, the activity of the L-arginine (L-*Arg*) nitric oxide (*NO*) pathway. The mechanism may involve inactivation of G_i proteins (G_i), decreased intracellular availibility of L-arginine and increased breakdown of NO by superoxide (O_2^-) oxLDL further activates endothelin (*ET*) gene expression and production. *SIN-1*, linsidomine; *BK*, bradykinin; *EDHF*, endothelium-derived hyperpolarizing factor. See legends to Figs. 1, 8 for other abbreviations

stimulus for the increased endothelin production are low-density lipoproteins which increase endothelin gene expression and release of the peptide from porcine and human aortic endothelial cells [75]. As these effects are specific for oxidized low-density lipoproteins and not shared by native low-density lipoproteins, it appears that the oxidation process as it occurs in atherosclerotic human arteries is crucial. In addition to endothelial cells, vascular smooth muscle cells, particularly those which migrated into the intima during the atherosclerotic process also produce endothelin. In cultured vascular smooth muscle cells, endothelin can be released by growth factors such as platelet-derived growth factor, transforming growth factor β_1 and vasoconstrictors such as arginine vasopressin [76]. Hence, several mediators involved in atherosclerosis do stimulate vascular endothelin production.This may explain why plasma endothelin levels are increased and are positively correlated with the extent of the atherosclerotic process [58]. Furthermore, particularly unstable lesions removed from coronary arteries by atherectomy do exhibit marked staining for endothelin-1 [77]. It is possible that local vascular endothelin contributes to abnormal coronary vasomotion in patients with unstable angina. Triggers of endothelin production in patients with acute coronary syndromes might be ischemia and thrombin. In atherosclerosis endothelin may contribute to hypervasoconstriction and proliferation of vascular smooth muscle cells.

Coronary Vasoconstriction and Spasm

In variant angina, coronary vasoconstriction is the sole or dominant mechanism of ischemia. In patients with obstructing coronary atherosclerosis, structural coronary obstruction per se limits increases in blood flow, but coronary vasoconstriction also importantly contributes (Fig. 12). The latter explains why the angina threshold varies on different occasions. Interestingly, in patients with variant angina, coronary spasm can be provoked by substances such as acetylcholine, histamine or ergonovine which normally evoke endothelium-dependent vasodilation. In isolated coronary arteries, acetylcholine, histamine and serotonin elicit endothelium-dependent relaxations due to stimulation of NO release [70, 73]; after removal of the endothelium the agonists exhibit marked contractions due to the presence of excitatory muscarinic, H_2-histaminergic and 5-HT_2-serotonergic receptors on vascular smooth muscle. Hence, abnormal coronary responses to acetylcholine, histamine, serotonin and ergonovine can be explained by endothelial dysfuntion. Patients with atherosclerotic coronary artery disease also exhibit paradoxical coronary vasoconstriction to acetylcholine and serotonin, while mild vasodilation is observed in normal coronary arteries [57. 78].

The mechanisms of this endothelial dysfunction is uncertain, but may involve effects of risk factors (see above) and alterations in signal transduction mechanisms. In porcine coronary arteries, in vivo endothelium denudation is followed by rapid endothelial regeneration and restoration of endothelium-dependent relaxations [79]. However, after several weeks, the number of endothelial cells per given area is increased and endothelium-dependent relaxations to serotonin, α_2-adrenergic agonists and ergonovine and aggregating platelets are reduced, while those to bradykinin, adenosine diphosphate or thrombin are maintained. As in atherosclerosis a defective release of NO due to dysfunction of G_i protein-coupled receptors with overwhelming contractile effects on vascular smooth muscle has been suggested.

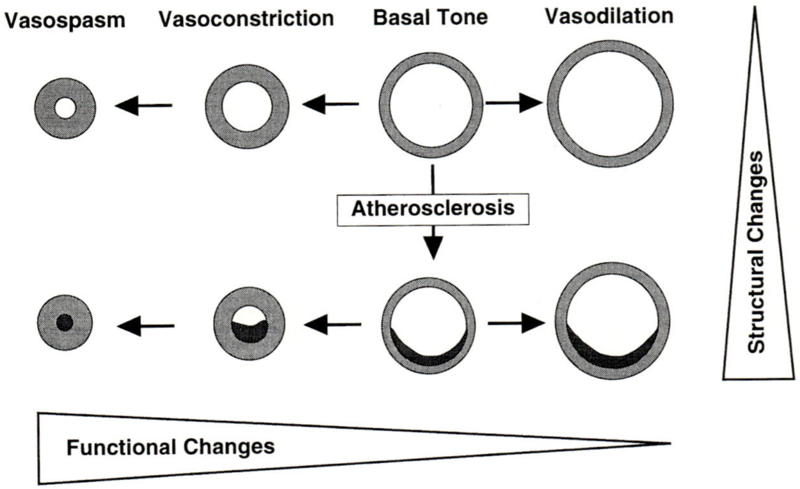

Fig. 12. Structural and functional changes in the vascular wall

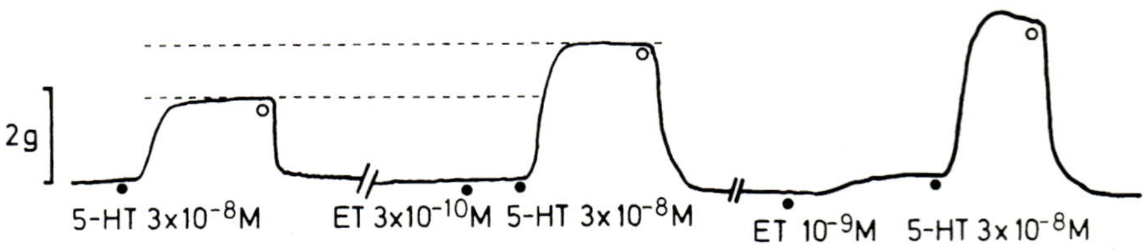

Fig. 13. Potentiating effects of endothelin (*ET*) on serotonin (*5-HT*)-induced contractions in human coronary arteries. (Modified from [80])

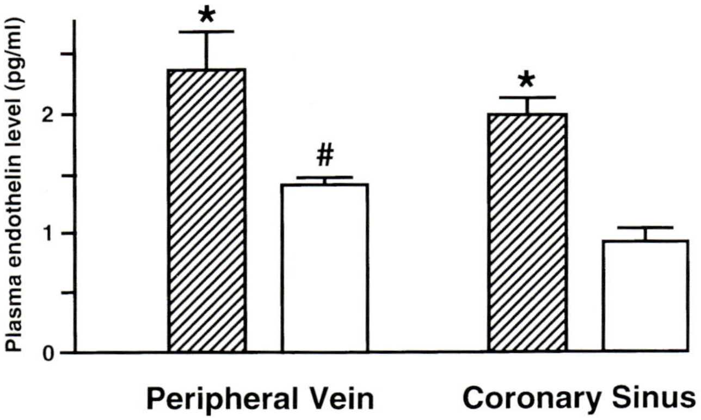

Fig. 14. Endothelin plasma levels in peripheral vein and coronary sinus in patients with variant angina (*hatched columns*) as compared to controls (*open columns*) (* $p < 0.05$ vs controls, $p < 0.05$ vs coronary sinus; modified from [81])

Endothelin-1 with its capacity to cause profound coronary vasoconstriction (see above) could also be involved in coronary spasm. However, the time course of its response is slow, but sustained and does not resemble the rapid vasoconstriction in patients. However, endothelin – particularly at low and threshold concentrations at which its own contractile effects are absent or negligible – increases the sensitivity of vascular smooth muscle to calcium and potentiates coronary contractions (Fig. 13; [80]). In patients with variant angina, endothelin levels in the coronary sinus are increased (Fig. 14; [81]). In contrast, in patients with clinical symptoms consistent with variant angina, but no provokable vasospasm, the levels remain within the normal range.

Restenosis

The treatment of choice of patients with atherosclerotic lesions of epicardial coronary arteries are either coronary bypass surgery or percutaneous transluminal angioplasty (PTCA). Both procedures do not provide definitive cure and are associated with bypass graft disease (see below) or restenosis respectively. PTCA leads to a more or less controlled injury of the blood vessel wall with deendothelialization, intimal dissection and as a consequence platelet adhesion and aggregation as well as thrombin formation. The presence of platelet-derived vasoconstrictors and growth factors as well as the direct mechanical stimulation of the wall leads to vascular smooth muscle proliferation and migration, particularly in the absence of endothelium-derived inhibitory factors (Fig. 8; [50, 82]). Endothelial cells regenerate thereafter and the lesion is again covered by endothelium. These cells, however, exhibit functional alterations, particularly as regards to the release of nitric oxide via G_i protein-coupled receptors (see above). Whether or not these cells also release increased amounts of growth factors is unproven, but likely. If this response to injury remains controlled, an excellent result of PTCA is achieved, while with excessive mirgration, proliferation and matrix formation restenosis ensues.

Function of Coronary Bypass Grafts

Bypass grafting with saphenous veins and internal mammary arteries are common practice to treat patients with coronary artery disease. The patency of mammary artery grafts however is superior to those of veins. Indeed, over 10 years after graft implantation mammary arteries exhibit a patency rate between 90% and 100%, while venous grafts are occluded in 50% of the cases [83]. Endothelial function differs considerably in the two vessels. Mammary arteries have pronounced endothelium-dependent relaxations to acetylcholine [9], bradykinin, thrombin and adenosine diphosphate. The latter agonist is responsible for endothelium-dependent relaxations to aggregating platelets in this artery (Fig. 15; [24]). In contrast in saphenous veins, endothelium-dependent relaxations to acetylcholine are weak and those to thrombin and aggregating platelets are absent [9, 25]. Rather, in response to platelets

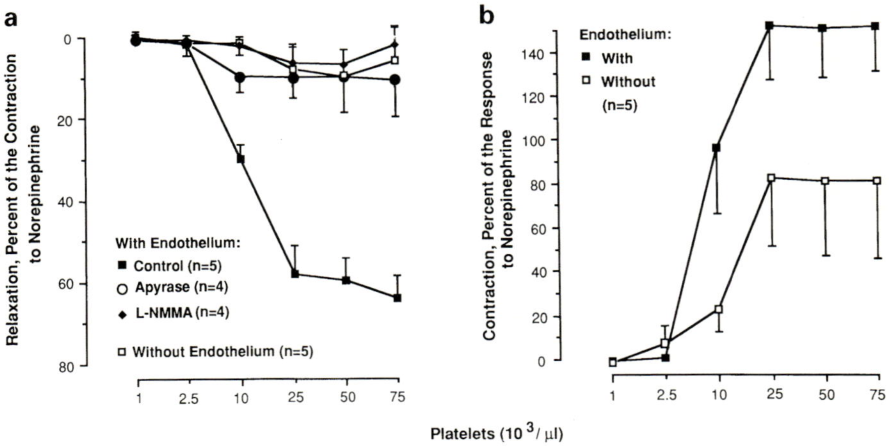

Fig. 15 a, b. Response to aggregating platelets in human internal mammery artery (a) and saphenous vein (b). Note the endothelium-dependent relaxations in the artery mediated by nitric oxide (due to stimulation by adenine nucleotides; apyrase is an enzyme which breaks down ADP and ATP). In the vein only contractions occur. (Modified from [24])

and thrombin marked contractions of saphenous veins occur (Fig. 15; [24]). In addition, growth factors released from aggragating platelets such as platelet-derived growth factor cause pronounced proliferation of vascular smooth muscle cells obtained from saphenous veins, but not from mammary arteries [24].

Furthermore, flow-dependent vasodilation is much more pronounced in mammary artery grafts than in venous grafts. In addition to a more pronounced nitric oxide release, mammary arteries also produce more prostacyclin than saphenous veins. On the other hand, vasoconstrictor responses to endothelin-1 are similar in the two blood vessels. These differences in the release of nitric oxide and prostacyclin to various mediators are likely to be of great importance for graft function and patency.

Pharmacological Modulation of Endothelium Function

Due to the importance of the endothelium in cardiovascular disease, endothelial effects of drugs which are currently used in clinical practice or in development are of great interest. Drugs can directly affect endothelial function, prevent the action of endothelial mediators, substitute for deficient endothelial factors or indirectly exert protective effects by interferering with cardiovascular risk factors.

Angiotensin Converting Enzyme Inhibitors

ACE inhibitors not only prevent activation of angiotensin I, but also the breakdown of bradykinin (Fig. 7; [84]). In the perfused porcine eye, the ACE inhibitor enalaprilat

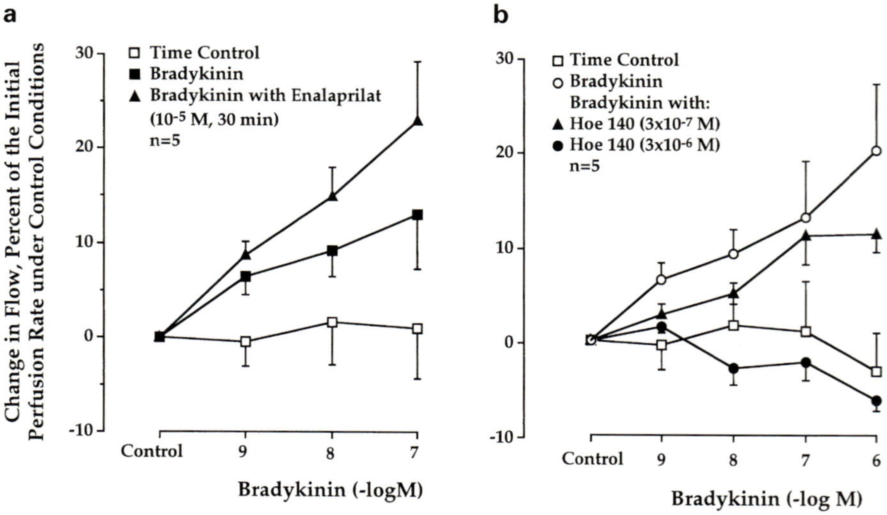

Fig. 16 a, b. Bradykinin-induced increase in ophthalmic flow of isolated porcine eyes. The respone is increased by the angiotensin converting enzyme inhibitor enalaprilat (**a**) and reduced by the B_2 bradykinin receptor antagonist Hoe 140 (**b**). (Modified from [85])

enhances the increase in local flow induced by bradykinin (which is mediated entirely by nitric oxide), while the B_2-bradykinin antagonist Hoe 140 prevents the response (Fig. 16; [85]). Similar observations have been made in human coronary arteries [86]. Hence, local vascular concentrations of bradykinin are importantly determined by ACE activity. Bradykinin is released by endothelial cells, possibly in response to shear forces, and stimulates NO and prostacyclin production in the endothelium via B_2-bradykinin receptors. ACE inhibitors therefore may increase flow-dependent vasodilation and the local formation of NO (H. Drexler, personal communication) and prostacyclin in patients. It is possible that the latter effects importantly contribute to the vascular protective effects of the drugs in patients after myocardial infarction, in particular to the reduction in acute coronary syndromes.

L-Arginine

A further approach to increase NO is supplementation with L-arginine, the precursor of the endothelial mediator [4]. In atherosclerotic rabbits, chronic dietary L-arginine supplementation improves endothelium-dependent relaxations and reduces the formation of fatty streaks [87]. In patients with coronary artery disease, intracoronary infusion of L-arginine does not ameliorate vasomotion of epicardial coronary arteries, but improves endothelium-dependent increases in coronary flow in the microcirculation [88]. The effects of chronic L-arginine supplementation in patients with coronary artery disease is still unknown.

Nitrovasodilators

Nitrates act by releasing NO from their molecule [89] which in turn activates soluble guanylyl cyclase and leads to the formation of cyclic GMP (Fig. 1; see above). Hence, nitrates can substitute for endogenous nitric oxide in disease states with a reduced formation of NO. In vessels without endothelium or after inhibition of NO formation, nitrovasodilators are paricularly effective (Fig. 17; [90]). Indeed in human arteries, the nitrovasodilator linsidomine is more effective after removal of the endothelium [90]. In the human forearm circulation, sodium nitroprusside exerts more pronounced increases in forearm blood flow after inhibition of NO production with L-NMMA [7]. With progressing atherosclerotic changes, the response of vascular smooth muscle to nitrates usually is reduced.

The potency of nitrovasodilators on a molar basis is related to the number of NO molecules they release[89]. Trinitrates release three, dinitrates two and mononitrates one NO molecule. Clinically, however, these differences are not important as the dosage of the compounds corrects for these differences. The potency of nitrates is reduced due to nitrate tolerance (i.e., the loss of vasodilator effects with prolonged application of the drugs). Tolerance can occur due to activation of the sympathetic nervous system and renin angiotensin system, decreased biotransformation of organic nitrates (due to a reduced availability of sulfhydryl groups and other mechanisms) and possibly a reduced activity of guanylyl cyclase [91]. Nitrovasodilators

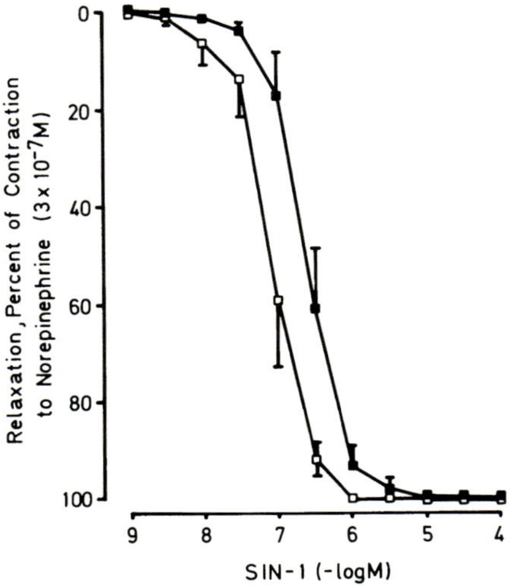

Fig. 17. Relaxations induced by the nitric oxide donor SIN-1 (the active metabolite of molsidomine). Note the increased response in vessels without endothelium. (Modified from [90]) *Open squares*, without endothelium ($n=6$); *filled spuares*, with endothelium ($n=8$)

which do not require enzymatic biotransformation and spontaneously release NO from their molecule (i.e., molsidomine) appear to be less prone for tolerance. A most recent finding in the rabbit aorta suggests that production of oxygen-derived free radicals by the endothelium and in turn inactivation of the NO released from the molecule of nitrovasodilators contributes importantly to tolerance.

Calcium Antagonists

Calcium channel blockers inhibit inflow of extracellular calcium through voltage-operated channels into the smooth muscle cells and thereby cause vasodilation (for details, see chapter by Wenzel et al., pp. 129–147). They have little effect on the endothelium, but facilitate the vasodilation induced by NO and reduce the contractile effects of endothelin-1. Furthermore, calcium channel blockers reduce proliferative responses of human coronary smooth muscle cells to platelet-derived growth factor, but not to mechanical stretch [82]. Angiographic trials have confirmed a modest antiatherosclerotic effect of calcium blockers in patients (for details see chapter by Wenzel et al., pp. 129–147).

Endothelin Receptor Antagonists

Endothelin antagonists are new compounds whitch specifically interfere with endothelin receptors (for details see chapter by Wenzel et al., pp. 129–147). Most substances are specific for ET_A-receptor, while newer molecules possess combined ET_A- and ET_B-receptor antagonism [46]. Certain of these compounds are orally active (in particular bosentan, SB209670 and PD 142893 [92, 93]). In isolated arteries including those obtained from humans, endothelin-induced vasoconstriction is potently inhibited by these drugs [94]. In the human skin microcirculation and the forearm circulation, ET_A-receptor antagonists increase blood flow and inhibit endothelin-induced vasoconstriction demonstrating the presence of endothelin-induced vascular tone in man [93, 95]. Phase I studies with certain compounds are under way.

Summary

The circulation is controlled by the central nervous system, circulating hormones and local vascular mechanisms. The importance of local vascular regulation has only recently been appreciated. The endothelium lies in a strategical anatomical position within the blood vessel wall between the circulating blood (and platelets and monocytes therein) and vascular smooth muscle cells. Endothelial cells are stimulated by mechanical and hormonal signals. In response to these stimuli it releases mediators modulating contraction and proliferation of vascular smooth muscle, platelet adhesion and aggregation, coagulation and monocyte adhesion. Nitric oxide and prostacyclin and a putative hyperpolarizing factor mediate relaxation. Nitric oxide also inhibits smooth muscle proliferation and together with prostacyclin platelet adhesion and aggregation. Bradykinin induced nitric oxide production is regulated by angiotensin converting enzyme on the endothelial surface as this enzyme not only converts angiotensin I into angiotensin II, but also inactivates bradykinin. Endothelin-1, thromboxane A_2 and prostaglandin H_2 are endothelium-derived contracting factors. In contrast to thromboxane A_2 and prostaglandin H_2 which activate platelets, endothelin-1 has no direct effects on these cells, but has proliferative properties in vascular smooth muscle. Under physiological conditions, the endothelium plays a protective role as it prevents adhesion of circulating blood cells, keeps the vasculature in a vasodilated state and inhibits vascular smooth muscle proliferation. In disease states, however, endothelial dysfunction contributes to enhanced vasoconstriction, adhesion of platelets and monocytes and proliferation of vascular smooth muscle cells, all events which could contribute to the development of several forms of cardiovascular disease. Nitrates substitute in part for deficient endogenous nitric oxide, while angiotensin converting enzyme inhibitors increase the bradykinin induced nitric oxide and prostacyclin production. Calcium antagonists facilitate the action of nitric oxide and reduce that of endothelin-1. The newly developed endothelin antagonists allow to specifically interfere with the effects of endothelin. Pharmacological correction of endothelial dysfunction may be important to treat cardiovascular disease and its complications.

Acknowledgment: Original research work of the authors reported in this article were supported by grants of the Swiss National Research Foundation (32-032541.91/2, 32-32655.91, SCORE 32-35591.92) and the Swiss Heart Foundation.

References

1. Lüscher TF, Vanhoutte PM (1990) The endothelium: modulator of cardiovascular function. CRC Press, Boca Raton
2. Furchgott RF, Zawadzki JV (1980) The obligatory role of endothelial cells in the relaxation of arterial smooth muscle by acetylcholine. Nature 299:373–376
3. Palmer RM, Ferrige AG, Moncada S (1987) Nitric oxide release accounts for the biological activity of endothelium-derived relaxing factor. Nature 327:524–526
4. Palmer RM, Ashton DS, Moncada S (1988) Vascular endothelial cells synthesize nitric oxide from L-arginine. Nature 333:664-666
5. Bredt DS, Hwang PM, Snyder SH (1990) Localization of nitric oxide synthase indicating a neural role for nitric oxide. Nature 347:768-770
6. Vallance P, Leone A, Calver A, Collier J, Moncada S (1992) Accumulation of an endogenous inhibitor of nitric oxide synthesis in chronic renal failure. Lancet 339:572-575
7. Joannides R, Haefeli WE, Linder L, Richard V, Bakkali EH, Thuillez C, Lüscher TF (1995) Nitric oxide is responsible for flow-dependent dilation of human peripheral conduit arteries in vivo. Circulation 91:1314-1319
8. Rees DD, Palmer RM, Schulz R, Hodson HF, Moncada S (1990) Characterization of three inhibitors of endothelial nitric oxide synthase in vitro and in vivo. Br J Pharmacol 101:746-752
9. Yang ZH, von Segesser L, Bauer E, Stulz P, Turina M, Lüscher TF (1991) Different activation of the endothelial L-arginine and cyclooxygenase pathway in the human internal mammary artery and saphenous vein. Circ Res 68:52-60
10. Tschudi M, Richard V, Bühler FR, Lüscher TF (1991) Importance of endothelium-derived nitric oxide in porcine coronary resistance arteries. Am J Physiol 260:H13–H20
11. Smith RE, Palmer RM, Bucknall CA, Moncada S (1992) Role of nitric oxide synthesis in the regulation of coronary vascular tone in the isolated perfused rabbit heart. Cardiovasc Res 26: 508-512
12. Meyer P, Flammer J, Lüscher TF (1993) Endothelium-dependent regulation of the ophthalmic microcirculation in the perfused porcine eye: role of nitric oxide and endothelins. Invest Ophthalmol Vis Sci 34:3614-3621
13. Vallance P, Collier J, Moncada S (1990) Effects of endothelium-derived nitric oxide on peripheral arteriolar tone in man. Lancet II:997-1000
14. Rees DD, Palmer RMJ, Moncada S (1989) Role of endothelium-derived nitric oxide in the regulation of blood pressure. Proc Natl Acad Sci USA 86:3375-3378
15. Arnal JF, Warin L, Michel JB (1992) Determinants of aortic cyclic guanosine monophosphate in hypertension induced by chronic inhibition of nitric oxide synthase. J Clin Invest 90:647–652
16. Margulies KB, Hildebrand FJ, Lerman A, Perrella MA, Burnett JJ (1990) Increased endothelin in experimental heart failure. Circulation 82:2226-2230
17. Moreau P, Takase H, Küng CF, van Rouijen M-M, Schaffner T, Lüscher TF (1995) Structure and function of the rat basilar artery during chronic nitric oxide synthase inhibition. Stroke (in press)
18. Rapoport RM, Draznin MB, Murad F (1983) Endothelium-dependent relaxation in rat aorta may be mediated through cyclic GMP-dependent protein phosphorylation. Nature 306:174-176
19. Stamler JS, Jaraki O, Osborne J, Simon Dl, Keaney J, Vita J, Singel D, Valeri CR, Loscalzo J (1992) Nitric oxide circulates in mammalian plasma primarily as an S-nitroso adduct of serum albumin. Proc Natl Acad Sci USA 89:7674–7677
20. Radomski MW, Palmer RM, Moncada S (1987) Comparative pharmacology of endothelium-derived relaxing factor, nitric oxide and prostacyclin in platelets. Br J Pharmacol 92:181–187
21. Radomski MW, Palmer RM, Moncada S (1987) Endogenous nitric oxide inhibits human platelet adhesion to vascular endothelium. Lancet 2:1057–1058
22. Radomski MW, Palmer RM, Moncada S (1990) An L-arginine/nitric oxide pathway present in human platelets regulates aggregation. Proc Natl Acad Sci USA 87:5193–5197
23. Cohen RA, Shepherd JT, Vanhoutte PM (1983) Inhibitory role of the endothelium in the response of isolated coronary arteries to platelets. Science 221:273–274
24. Yang Z, Stulz P, Von SL, Bauer E, Turina M, Lüscher TF (1991) Different interactions of platelets with arterial and venous coronary bypass vessels. Lancet 337:939–943

25. Lüscher TF, Diederich D, Siebenmann R, Lehmann K, Stulz P, von SL, Yang ZH, Turina M, Grädel E, Weber E, Bühler FR (1988) Difference between endothelium dependent relaxation in arterial and in venous coronary bypass grafts. N Engl J Med 319:462–467
26. Wright CE, Rees DD, Moncada S (1992) Protective and pathological roles of nitric oxide in endotoxin shock. Cardiovasc Res 26:48–57
27. Petros A, Bennett D, Vallance P (1991) Effect of nitric oxide synthase inhibitors on hypotension in patients with septic shock. Lancet 338:1557–1558
28. Moncada S, Vane VR (1979) Pharmacology and endogenous roles of prostaglandin endoperoxides, thromboxane A2 and prostacyclin. Pharmacol Rev 30:293–331
29. Richard V, Tanner FC, Tschudi M, Lüscher TF (1990) Different activation of L-arginine pathway by bradykinin, serotonin, and clonidine in coronary arteries. Am J Physiol 259:H1433–H1439
30. Vanhoutte PM (1987) Vascular physiology: the end of the quest? [news]. Nature 327:459–460
31. De Mey JG, Vanhoutte PM (1982) Heterogeneous behavior of the canine arterial and venous wall. Importance of the endothelium. Circ Res 51:439–447
32. Yanagisawa M, Kurihara H, Kimura S, Tomobe Y, Kobayashi M, Mitsui Y, Yazaki Y, Goto K, Masaki T (1988) A novel potent vasoconstrictor peptide produced by vascular endothelial cells. Nature 332:411 –415
33. Ohnaka K, Takayanagi R, Nishikawa M, Haji M, Nawata H (1993) Purification and characterization of a phosphoramidon-sensitive endothelin-converting enzyme in porcine aortic endothelium. OFF. J Biol Chem 268:26759–26766
34. Xu D, Emoto N, Giaid A, Slaughter C, Kaw S, de Wit D, Yanagisawa M (1994) ECE-1: a membrane-bound metalloprotease that catalyzes the proteolytic activation of big endothelin-1. Cell 78:473–485
35. Boulanger C, Lüscher TF (1990) Release of endothelin from the porcine aorta. Inhibition of endothelium-derived nitric oxide. J Clin Invest 85:587–590
36. Goerre S, Wenk M, Bärtsch P, Lüscher TF, Niroomand S, Hohernhaus H, Oelz O, Reinhart WH (1995) Endothelin-1 in pulmonary hypertension associated with high altitude. Circulation 91: 359–364
37. Kiowski W, Lüscher TF, Linder L, Bühler FR (1991) Endothelin-1-induced vasoconstriction in humans. Reversal by calcium channel blockade but not by nitrovasodilators or endothelium-derived relaxing factor. Circulation 83:469–475
38. Küng CF, Tschudi MR, Noll G, Clozel J-P, Lüscher TF (1995) Differential effects of calcium antagonist mibefradil in epicardial and intramyocardial coronary arteries. J Cardiovasc Pharmacol (in press)
39. Lüscher TF, Yang Z, Tschudi M, Von SL, Stulz P, Boulanger C, Siebenmann R, Turina M, Bühler FR (1990) Interaction between endothelin-1 and endothelium-derived relaxing factor in human arteries and veins. Circ Res 66:1088–1094
40. Lüscher TF, Boulanger CM, Dohi Y, Yang ZH (1992) Endothelium-derived contracting factors. Hypertension 19:117–130
41. Wagner OF, Christ G, Wojta J, Vierhapper H, Parzer S, Nowotny PJ, Schneider B, Waldhäusl W, Binder BR (1992) Polar secretion of endothelin-1 by cultured endothelial cells. J Biol Chem 267: 16066–16068
42. Yokokawa K, Kohno M, Yasunari K, Murakawa K, Takeda T (1991) Endothelin-3 regulates endothelin-1 production in cultured human endothelial cells. Hypertension 18:304–315
43. Stewart DJ, Langleben D, Cernacek P, Cianflone K (1990) Endothelin release is inhibited by coculture of endothelial cells with cells of vascular media. Am J Physiol 259:H1928–H1932
44. Arai H, Hori S, Aramori I, Ohkubo H, Nakanishi S (1990) Cloning and expression of a cDNA encoding an endothelin receptor. Nature 348:730–732
45. Sakurai T, Yanagisawa M, Takuwa Y, Miyazaki H, Kimura S, Goto K, Masaki TK (1990) Cloning of a cDNA encoding a non-isopeptide-selective subtype of the endothelin receptor. Nature 348: 732–735
46. Lüscher TF (1993) Do we need endothelin antagonists? Cardiovasc Res 27:2089–2093
47. Ng KK, Vane JR (1967) Conversion of angiotensin I to angiotensin II. Nature 216:762–766
48. Baumgartner HR, Studer A (1963) Gezielte Überdehnung der Aorta abdominalis am normo- und hypercholesterinaemischen Kaninchen. Pathol Microbiol (Basel) 26:129–148
49. Radomski MW, Palmer RM, Moncada S (1987) The anti-aggregating properties of vascular endothelium: interactions between prostacyclin and nitric oxide. Br J Pharmacol 92:639–646

50. Ross R (1993) The pathogenesis of atherosclerosis: a perspective for the 1990s. Nature 362: 801–809
51. Garg UC, Hassid A (1989) Nitric oxide-generating vasodilators and 8-bromo-cyclic guanosine monophosphate inhibit mitogenesis and proliferation of cultured rat vascular smooth muscle cells. J Clin Invest 83: 1774–1777
52. Battegay EJ, Raines EW, Seifert RA, Bowen-Pope DF, Ross R (1990) TGF-beta induces bimodal proliferation of connective tissue cells via complex control of an autocrine PDGF loop. Cell 63: 515–524
53. Castellot JJ Jr, Beeler DL, Rosenberg RD, Karnovsky MJ (1984) Structural determinants of the capacity of heparin to inhibit the proliferation of vascular smooth muscle cells. J Cell Physiol 120: 315–320
54. Hannan RL, Kourembanas S, Flanders KC, Rogelj SJ, Roberts AB, Faller DV, Klagsbrun M (1988) Endothelial cells synthesize basic fibroblast growth factor and transforming growth factor beta. Growth Factors 1: 7–17
55. DiCorleto PE, Hassid A (1990) Growth factor production by endothelial cells. In: Ryan U (ed) Endothelial cells. CRC Press, Boca Raton, pp 51–62
56. Küng CF, Lüscher TF (1995) Different mechanisms of endothelial dysfunction with aging and hypertension in rat aorta. Hypertension 25: 194–200
57. Zeiher AM, Drexler H, Saurbier B, Just H (1993) Endothelium-mediated coronary blood flow modulation in humans. Effects of age, atherosclerosis, hypercholesterolemia, and hypertension. J Clin Invest 92: 652–662
58. Lerman A, Edwards BS, Hallett JW, Heublein DM, Sandberg SM, Burnett JJ (1991) Circulating and tissue endothelin immunoreactivity in advanced atherosclerosis. N Engl J Med 325: 997–1001
59. Dohi Y, Criscione L, Lüscher TF (1991) Renovascular hypertension impairs formation of endothelium-derived relaxing factors and sensitivity to endothelin-1 in resistance arteries. Br J Pharmacol 104: 349–354
60 Lüscher TF (1992) Heterogeneity of endothelial dysfunction in hypertension. Eur Heart J 13 [Suppl D]: 50–55
61. Lüscher TF (1994) The endothelium in hypertension: bystander, target or mediator? J Hypertens 12 [Suppl 10]: S105–S116
62. Linder L, Kiowsky W, Bühler FR, Lüscher TF (1990) Indirect evidence for release of endothelium-derived relaxing factor in human forearm circulation in vivo. Blunted response in essential hypertension. Circulation 81: 1762–1767
63. Panza JA, Quyyumi AA, Brush JJ, Epstein SE (1990) Abnormal endothelium-dependent vascular relaxation in patients with essential hypertension. N Engl J Med 323: 22–27
64. Cockcroft JR, Chowienczyk PJ, Benjamin N, Ritter JM (1994) Preserved endothelium dependent vasodilatation in patients with essential hypertension. N Engl J Med 330: 1036–1040
65. Nava E, Noll G, Lüscher TF (1995) Increased cardiac activity of nitric oxide synthase in spontaneous hypertension. Circulation 91: 2310–2313
66. Taddei S, Virdis A, Mattei P, Salvetti A (1993) Vasodilation to acetylcholine in primary and secondary forms of human hypertension. Hypertension 21: 929–933
67. Lüscher TF, Vanhoutte PM, Raij L (1987) Antihypertensive treatment normalizes decreased endothelium-dependent relaxations in rats with salt-induced hypertension. Hypertension 9 [Suppl III]: 193–197
68. Larivière R, Thibault G, Schiffrin EL (1993) Increased endothelin-1 content in blood vessels of deoxycorticosterone acetate-salt hypertensive but not in spontaneously hypertensive rats. Hypertension 21: 294–300
69. Harrison DG, Freiman PC, Armstrong ML, Marcus ML, Heistad DD (1987) Alterations of vascular reactivity in atherosclerosis. Circ Res 61: 1174–1180
70. Tanner FC, Noll G, Boulanger CM, Lüscher TF (1991) Oxidized low density lipoproteins inhibit relaxations of porcine coronary arteries. Role of scavenger receptor and endothelium-derived nitric oxide. Circulation 83: 2012–2020
71. Shimokawa H, Flavahan NA, Vanhoutte PM (1989) Natural course of the impairment of endothelium-dependent relaxations after balloon endothelium removal in porcine coronary arteries. Possible dysfunction of a pertussis toxin-sensitive G protein. Circ Res 65: 740–753
72. Minor R Jr, Myers PR, Guerra R Jr, Bates JN, Harrison DG (1990) Diet-induced atherosclerosis increases the release of nitrogen oxides from rabbit aorta. J Clin Invest 86: 2109–2116

73. Shimokawa H, Vanhoutte PM (1989) Impaired endothelium-dependent relaxation to aggregating platelets and related vasoactive substances in porcine coronary arteries in hypercholesterolemia and atherosclerosis. Circ Res 64:900–914
74. Winkles JA, Alberts GF, Brogi E, Libby P (1993) Endothelin-1 and endothelin receptor mRNA expression in normal and atherosclerotic human arteries. Biochem Biophys Res Commun 191: 1081–1088
75. Boulanger CM, Tanner FC, Bea ML, Hahn AW, Werner A, Lüscher TF (1992) Oxidized low density lipoproteins induce mRNA expression and release of endothelin from human and porcine endothelium. Circ Res 70:1191–1197
76. Hahn AW, Resink TJ, Scott-Burden T, Powell J, Dohi Y, Bühler FR (1990) Stimulation of endothelin mRNA and secretion in rat vascular smooth muscle cells: a novel autocrine function. Cell Regul 1:649–659
77. Zeiher AM, Ihling C, Pistorius K, Schächinger V, Schaefer H-E (1994) Increased tissue endothelin immunoreactivity in atherosclerotic lesions associated with acute coronary syndromes. Lancet 344:1405–1406
78. Ludmer PL, Selwyn AP, Shook TL, Wayne RR, Mudge GH, Alexander RW, Ganz P (1986) Paradoxical vasoconstriction induced by acetylcholine in atherosclerotic coronary arteries. N Engl J Med 315:1046–1051
79. Shimokawa H, Aarhus LL, Vanhoutte PM (1987) Porcine coronary arteries with regenerated endothelium have a reduced endothelium-dependent responsiveness to aggregating platelets and serotonin. Circ Res 61:256–270
80. Yang Z, Richard V, von Segesser L, Bauer E, Stulz P, Turina M, Lüscher TF (1990) Threshold concentrations of endothelin-1 potentiate contractions to norepinephrine and serotonin in human arteries. A new mechanism of vasospasm? Circulation 82:188–195
81. Toyo-oka T, Aizawa T, Suzuki N, Hirata Y, Miyauchi T, Shin WS, Yanagisawa M, Masaki T, Sugimoto T (1991) Increased plasma level of endothelin-1 and coronary spasm induction in patients with vasospastic angina pectoris. Circulation 83:476–483
82. Yang Z, Noll G, Lüscher TF (1993) Calcium antagonists differently inhibit proliferation of human coronary smooth muscle cells in response to pulsatile stretch and platelet-derived growth factor. Circulation 88:832–836
83. Loop FD, Lytle BW, Cosgrove DM, Stewart RW, Goormastic M, Williams GW, Golding LA, Gill CC, Taylor PC, Sheldon WC et al (1986) Influence of the internal-mammary-artery graft on 10-year survival and other cardiac events. N Engl J Med 314:1–6
84. Wiemer G, Scholkens BA, Becker RH, Busse R (1991) Ramiprilat enhances endothelial autacoid formation by, inhibiting breakdown of endothelium-derived bradykinin. Hypertension 18: 558–563
85. Meyer P, Flammer J, Lüscher TF (1995) Local action of the renin angiotensin system in the porcine ophthalmic circulation: effects of ACE-inhibitors and angiotensin receptor antagonists. Invest Ophthalmol Vis Sci (in press)
86. Auch-Schwelk W, Bossaller C, Claus M, Graf K, Grafe M, Fleck E (1993) ACE inhibitors are endothelium dependent vasodilators of coronary arteries during submaximal stimulation with bradykinin. Cardiovasc Res 27:312–317
87. Cooke JP, Singer AH, Tsao P, Zera P, Rowan RA, Billingham ME (1992) Antiatherogenic effects of L-arginine in the hypercholesterolemic rabbit. J Clin Invest 90:1168–1172
88. Drexler H, Zeiher AM, Meinzer K, Just H (1991) Correction of endothelial dysfunction in coronary microcirculation of hypercholesterolaemic patients by L-arginine. Lancet 338:1546–1550
89. Feelisch M, Noack EA (1987) Correlation between nitric oxide formation during degradation of organic nitrates and activation of guanylate cyclase. Eur J Pharmacol 139:19–30
90. Lüscher TF, Richard V, Yang ZH (1989) Interaction between endothelium-derived nitric oxide and SIN-1 in human and porcine blood vessels. J Cardiovasc Pharmacol 14 [Suppl 11]: S76–S80
91. Abrams J (1991) Clinical aspects of nitrate tolerance. Eur Heart J 12 [Suppl E]: 42–52
92. Clozel M, Breu V, Burri K, Cassal J-M, Fischli W, Gray GA, Hirth G, Löffler B-M, Müller M, Neidhart W, Ramuz H (1993) Pathophysiologic role of endothelin revealed by the first orally active endothelin receptor antagonist. Nature 365:759–761
93. Wenzel RR, Noll G, Lüscher TF (1994) Endothelin receptor antagonists inhibit endothelin in human skin microcirculation. Hypertension 23:581–586

94. Seo B, Oemar BS, Siebenmann R, von Segesser L, Lüscher TF (1994) Both ETA and ETB receptors mediate contraction to endothelin-1 in human blood vessels. Circulation 89:1203–1208
95. Haynes WG, Webb DJ (1994) Contribution of endogenous generation of endothelin-1 to basal vascular tone. Lancet 344:852–854

Effects of Physical and Pharmacological Interventions on Circulating Endothelin-1 Levels in Humans

A. Morganti, F. Ghio, and G. Gazzano

Introduction

Endothelin-1 (ET-1) is a 21 amino acid peptide produced by the endothelial cells that for its potent and long lasting vasoconstrictive action may be involved in a number of cardiovascular disorders [1, 2]. As to the mechanisms of regulation of ET-1, "in vitro" studies have shown that its secretion is controlled by numerous physical and humoral factors (Table 1); however "in vivo" studies addressed at investigating the physiological relevance of these factors may encounter difficulties for at least two reasons. One is that under normal conditions the concentration of ET-1 in blood is low (between 1–2 pg/ml); thus, since its secretion occurs by 80% or more towards the underlying vascular smooth muscle cells [3, 4], the changes in plasma ET-1 in response to stimuli may be too small to be detected with confidence with the present radioimmunoassays. The second reason is that there is no intracellular storage of ET-1 in most tissues [5, 6]; therefore, in order to observe modifications of the circulating peptide the duration of the stimulus must be long enough to activate "de novo" its synthesis. These reasons probably explain why most studies aimed at elevating plasma ET-1 have been inconclusive so far.

In this paper we will briefly report the results of some studies of ours addressed at investigating in man the effects of various physical stimuli which may potentially modify ET-1 secretion; the effects of pharmacological manipulations with calcium antagonists and angiotension converting enzyme (ACE) inhibitors were also examined.

Table 1. Agents affecting endothelin secretion

Stimulators			Suppressors
Physical	Hormones	Others	
Shear stress	Vasopressin	Thrombin	Nitric oxide
Hypoxia	Angiotensin II	TGF	ANF
	Thromboxane A2	LDL	Cyclic GMP
	Epinephrine	Endotoxin	
	Insulin	Ca^{2+} ionophores	
	Glucocorticoids	Phorbol esters	
	Bradykinin	Interleukin-1	

TGF, transforming growth factor; LDL, low density lipoprotein; ANF, atrial natriuretic factor

High Altitude Hypoxia

There is abundant evidence that hypoxia is a potent stimulus for ET-1 synthesis and release; indeed it has been shown that hypoxia induces endothelin gene expression and secretion in cultured rat and human endothelium [7, 8] and causes, in animal preparations, an endothelium dependent vasoconstriction that is associated with increased ET-1 secretion [9]. However, the effects of hypoxia on ET-1 in man are still largely unknown. To clarify this we recently performed two twin studies in which normal subjects were exposed to high altitude hypoxia [10]. In one study eight normal subjects ascended over a period of 8 days from sea level to the altitude of 5050 m on Mount Everest in Nepal where they rested for 6 additional days; in a second study ten other normal subjects were exposed to a similar altitude but at a different latitude, i.e., at the 4559 m of Capanna Margherita on Mount Rosa in the western Italian Alps where they sojourned for two days. Venous blood samples for ET-1 measurements were collected at sea level before departure, during the sojourn at high altitude and, in the Mount Everest study, also on several occasions along the route. In this study we found that plasma ET-1 progressively and significantly increased from 1.8 ± 0.1 pg/ml at sea level to 2.7 ± 0.2 pg/ml at the altitude of 4240 m while oxygen saturation specularly declined from $99 \pm 0.2\%$ to $81 \pm 0.4\%$; both ET-1 and oxygen saturation leveled off thereafter and, overall, their changes were inversely correlated ($r = 0.5$, $p < 0.01$).

In the Capanna Margherita study plasma ET-1 rose almost to the same level observed in the Mount Everest study, i.e., from 1.7 ± 0.1 pg/ml to 2.3 ± 0.2 pg/ml and oxygen saturation fell from $99 \pm 0.2\%$ to $85 \pm 0.6\%$; again , these changes were statistically significant ($p < 0.05$ for both) and inversely correlated. In addition, in this study the changes in plasma ET-1 were also correlated with those in systolic pulmonary pressure (SPP, r=0.78, p<0.05) which was evaluated by Doppler echocardiography and rose on average from 19 ± 1 to 26 ± 2 mmHg.

The results of these studies go along with those recently reported by Goerre et al. in mountaineers who were examined in the same setting of our investigations [11] and confirm that hypobaric hypoxia can actually increase plasma ET-1 in normal subjects; moreover, the correlation between these increments and those of SPP suggest that under these circumstances the peptide can contribute to the adaptation of pulmonary circulation to the stress of high altitude. The possibility that ET-1 has this physiological function is supported by the observation that, in rats, chronic blockade of ET-1 receptors with specific antagonists prevents hypoxia induced pulmonary hypertension [12] and also with the finding that plasma levels of ET-1 similar to those observed in this study were seen in patients with primary or secondary pulmonary hypertension [13, 14] or chronically hypoxemic [15].

Cold Exposure

There are at least two reports in the literature showing that immersion of one forearm into icewater for 2 min (cold pressor test, CPT) induces a rapid increase in ET-1 in venous blood samples collected from the immersed and nonimmersed arm; these observations were made in normal subjects [16] and in patients with Raynaud's phenomenon [17], these latter having baseline plasma ET-1 levels higher than normal. These findings are of particular interest because, at variance with the lack of intracellular storage mentioned above, suggest that there are tissues from where ET-1 can be rapidly mobilized. We also examined the effects of CPT in patients with uncomplicated hypertension in whom baseline ET-1 was between 1 and 2 pg/ml but we failed to find consistent changes of the peptide in blood samples collected from both arms (Fig. 1) [18]. Equally negative studies were obtained when CPT was administered after patients were given a 30 mg tablet of nifedipine orally (Fig. 2); interestingly, also baseline values of ET-1 were unaffected by nifedipine administration. Thus, from our results it appears that a brief exposure to cold, alone or in combination with nifedipine and the attendant reduction in blood pressure, does not modify ET-1 secretion; these negative findings are inagreement with those of Hassi et al. [19] who equally failed to find significant changes of ET-1 in man during a 2-h stay in a cold room. It is still possible that in patients with high baseline values of ET-1 the responsiveness of the peptide to cold may be greater. To test this possibility we are currently examining

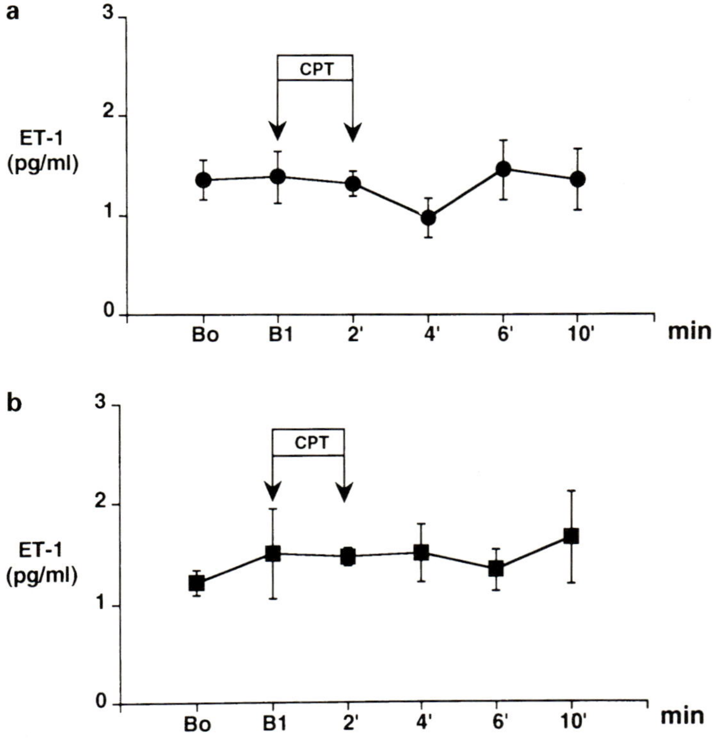

Fig. 1a, b. Effects of cold pressor text (*CPT*) on endothelin-1 (*ET-1*) in patients with essential hypertension. Patients were off antihypertensive treatment at the time of the study. Blood for ET-1 measurement were collected twice (B_o, B_1) from a peripheral vein before CPT from the immersed arm (**a**) and from the contralateral arm (**b**) and again at the time intervals shown on the horizontal axis. (From [18])

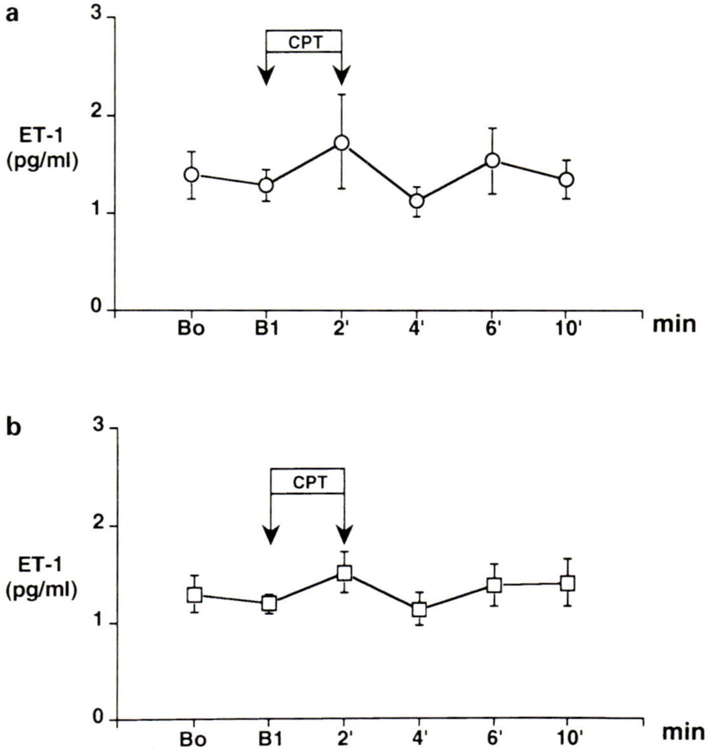

Fig. 2a, b. Effect of cold pressor test (*CPT*) on endothelin-1 (*ET-1*) in the same patients as in Fig. 1. Patients were given 30 mg nifedipine orally 3–4 h before the application of CPT. Blood was collected from the immersed (**a**) and contralateral arms (**b**) at the same time intervals as in Fig. 1. (From [18])

the response to CPT in patients with hypertension and mild renal insufficiency, a condition known to be associated with higher plasma levels of ET-1 [20, 21].

Postural Studies

In normal subjects the assumption of upright posture, either active or passive with tilt, has been shown to be associated with increments in ET-1 plasma levels [22, 23]; these findings suggest the possibility of a baroreceptor mediated regulation of ET-1 secretion because its increases were accompanied by those, in the same direction, of renin whose response to the postural stimulus is notoriously induced through a reflex sympathetic activation. We have been trying to reproduce these findings in patients with essential hypertension and baseline values of ET-1 scattered between 2 and 6 pg/ml; however in response to one hour of active standing we were unable to find consistent changes of ET-1 in peripheral venous blood whereas plasma renin activity rose in all subjects and, on average, from 2.6 ± 0.6 to 6.6 ± 1.3 ng/ml per hour (Fig. 3). We cannot exclude that in patients with essential hypertension the response of ET-1

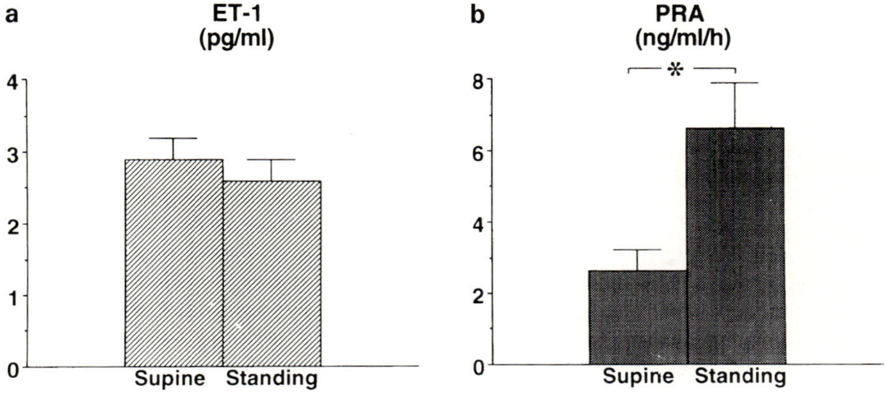

Fig. 3a, b. Effects of 1 h of active standing on plasma endothelin-1 (*ET-1*) and plasma renin activity (*PRA*) in 14 hypertensive patients. PRA consistently increased in all subjects

to upright posture may be blunted with respect to normal subjects, nor that the increments were missed because occurred earlier during standing; however, in general, caution must be applied in interpreting the results of these postural studies in that the changes in ET-1 may be due to modifications in the peptide clearance rather than to those in its secretion rate.

Alterations in the Activity of the Renin–Angiotensin System

Angiotensin II (A II) is a potent secretagouge of ET-1 in vitro [24]; thus, if this mechanism of regulation is physiologically relevant one may expect to see an association between levels of A II and those of ET-1 in vivo. To address this issue we recently investigated the relationship between A II and ET-1 in 12 patients with hypertension and unilateral renal artery stenosis who were subdivided according to their peripheral A II levels in those with high A II (i.e., >20 pg/ml, $n=7$) and in those with normal A II ($n=5$) [25]. In these studies blood samples for A II and ET-1 measurements were collected from the aorta (or alternatively from the inferior vena cava) and from both renal veins. We found that in patients with high A II, ET-1 was higher at each site of blood collection than in those with normal A II (respectively, 2.8 ± 0.3, 2.5 ± 0.2 and 2.4 ± 0.1 pg/ml vs 2.3 ± 0.4, 1.9 ± 0.4 and 1.9 ± 0.3 pg/ml, $p<0.05$); moreover, overall values of A II and of ET-1 were slightly but significantly correlated ($r=0.37$, $p<0.01$). Also, in seven patients (four with high A II and three with normal A II) who were restudied 60 min after a technically successful percutaneous transluminal angioplasty of renal artery stenosis, ET-1 was unchanged at all sites of blood collection in spite of concomitant clearcut reductions of A II levels which were particularly evident in samples collected in the vein of the kidney with the dilated artery (from 35 ± 25 to 22 ± 11 pg/ml, $p<0.05$). Also in another group of eight patients in whom we measured A II and ET-1 in peripheral blood throughout the 5 days after renal angioplasty we found that ET-1 was unchanged with respect to basal values (2.5 ± 0.2 pg/ml) in spite

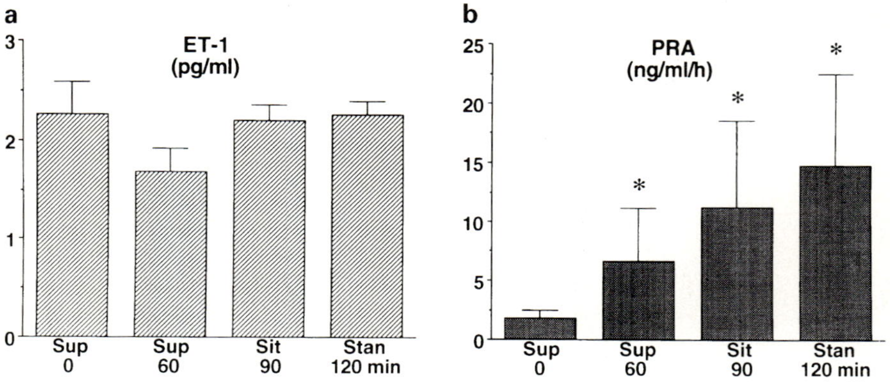

Fig. 4a, b. Acute effects of angiotension converting enzyme inhibition with 25 mg captopril given orally to five hypertensive patients. Minutes on the *horizontal axis* refers to time after the ingestion of captopril and *asterisks* refer to the statistical significance of the increments in plasma renin activity with respect to baseline values in the supine (*Sup*) position ($p < 0.05$). *Sit*, sitting; *Stan*, standing

of the fall of A II from 28 ± 11 to 4 ± 0.8 pg/ml. These results support the notion that A II may partecipate in ET-1 regulation in that abnormally high levels of A II as those occasionally found in some patients with renovascular hypertension may be associated with higher than normal values of ET-1; however, from our results it also appears that the stimulating action of A II on ET-1 must be exerted chronically to achieve an effect because acute or short-term reduction of A II is not accompanied by concomitant reduction in ET-1.

These results and their interpretation go along with those obtained using a different approach for reducing A II formation, i.e., with the administration of ACE inhibitors. Figure 4 illustrates how the ingestion of a single dose of 25 mg captopril in hypertensive patients examined in the supine, sitting and standing position failed to affect ET-1 levels whereas plasma renin activity increased already in the supine position, indicating the effective blockade of A II formation achieved with captopril. However, in another series of patients chronically treated with captopril or other ACE inhibitors and with almost undetectable plasma A II we found that circulating ET-1 had a tendency to be lower than in a comparable group of patients not treated with ACE inhibitors and obviously higher A II levels (Fig. 5).

Conclusions

From our studies it appears that among the various stimuli examined herein the only one which consistently affects plasma ET-1 was the prolonged exposure to high altitude. These changes in ET-1 are most likely related to hypobaric hypoxia and probably contribute to the adaptation of the pulmonary circulation to this condition. We also found a slight positive relationship between ET-1 and A II in patients with renovascular hypertension and a tendency for ET-1 to decrease in response to the chronic pharmacological blockade of A II formation; these observations are compatible with

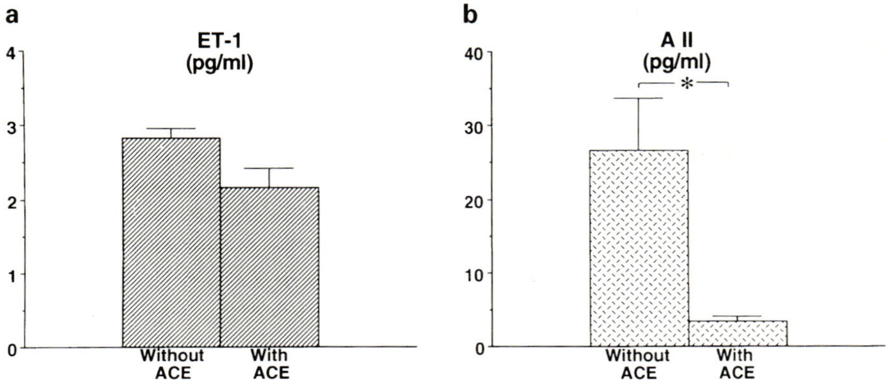

Fig. 5a, b. Comparison of endothelin-1 (*ET-1*; **a**) and angiotensin II (*A II*; **b**) levels in peripheral blood in two groups of hypertensive patients chronically treated with ($n=9$) and without angiotensin converting enzyme inhibitors ($n=19$). Note how in the former group A II values in plasma were very low. *Asterisk*, $p<0.05$

the possibility that A II is a physiological modulator of ET-1 secretion. However, the lack of changes in plasma ET-1 following the application of the other stimuli (cold, posture) does not necessarily exclude the possibility that they may have affected ET-1 production in tissues where this peptide may locally participate in the regulation of the vascular responses.

References

1. Haynes WG, Webb DJ (1993) The endothelin family of peptides: local hormones with diverse roles in health and disease? Clin Sci 84:485–500
2. Lüscher TF, Boulanger CM, Dohi Y, Yang Z (1992) Endothelium-derived contracting factors. Hypertension 19:117–130
3. Yoshimoto S, Ishizaki Y, Sasaki T, Murota SI (1991) Effect of carbon dioxide and oxygen on endothelin production by cultured porcine cerebral endothelial cells. Stroke 22:378–383
4. Wagner OF, Christ G, Woita J, Vierhapper H, Parzer S, Novotny PJ (1992) Polar secretion of endothelin-1 by cultured endothelial cells. J Biol Chem 267:16 066–16 068
5. Nakamura SM, Naruse M, Naruse K, Demura H, Uemura H (1990) Immunocytochemical localization of endothelin in cultured bovine cells. Histochemistry 94:475–477
6. Schini VB, Hendrickson H, Heublein DM, Burnett JC Jr, Vanhoutte PM (1989) Thrombin enhanced the release of endothelin from cultured porcine aortic endothelial cells. Eur J Pharmacol 165: 333–334
7. Elton TS, Oparil S, Taylor GR, Hicks PH, Yang RH, Jin H, Chen YF (1992) Normobaric hypoxia stimulates endothelin-1 gene expression in the rat. Am J Physiol 263:R1260–R1264
8. Kourembanas S, Marsden P, McQuillan L, Faller DV (1991) Hypoyia induces endothelin gene expression and secretion in cultured human endothelium. J Clin lnvest 88:1054–1057
9. Rakugi H, Tabuchi Y, Nakamura M, Nafano M, Higashimori K, Mikemi H (1990) Evidence of endothelin-1 release from resistance vessels of rats in response to hypoxia. Biochem Biophys Res Commun 169:973–977
10. Morganti A, Giussani M, Sala C, Gazzano G, Marana I, Pierini A, Savoia MT, Ghio F, Cogo A, Zanchetti A. Effects of exposure to high altitude on plasma ET-1 in normal man. J Hypertens (in press)

11. Goerre S, Wenk M, Bartsch P, Lüscher TF, Niroomand F, Hohenhaus E, Oelz O, Reinhart WH (1995) Endothelin-1 in pulmonary hypertension associated with high altitude exposure. Circulation 90:359–364
12. Chen SJ, Chen YF, Meng QC, Elton TS, Yano M, Oparil S (1994) Chronic blockade of endothelin-A receptors with BQ-123 prevents hypoxia induced pulmonary hypertension in the rat. Am J Hypertens 7 (2):96A, abstract H32
13. Stewart DJ, Levy RD, Cemacek P, Langleben D (1991) Increased plasma endothelin-1 in pulmonary hypertension: marker or mediator of disease. Ann Intern Med 114:464–469
14. Cody RJ, Haas GJ, Binkley PF, Capers Q, Kelly R (1992) Plasma endothelin correlates with the extent of pulmonary hypertension in patients with chronic congestive heart failure. Circulation 85:504–509
15. Ferri C, Bellini C, De Angelis C, Perrone A, De Siati C, Di Francesco L, Properzi G, Santucci A, Balsano F (1994) Circulating endothelin-1 levels in chronic hypoxemic patients. Am J Hypertens 7 (2): 135A, abstract L4
16. Fyhrquist F, Saijonmaa O, Messarinne K, Tikkanen I, Rosenlof K, Tikkanen T (1990) Raised plasma endothelin-1 concentration following cold pressor test. Biochem Biophys Res Commun 169: 217–221
17. Zamora MR, O'Brien RF, Rutherford RB, Weil JV (1990) Serum endothelin-1 concentrations and cold provocation in primary Raynaud's phenomenon. Lancet 336:1144–1147
18. Morganti A, Giussani M, Ghio F, Pierini A, Savoia MT, Cogo A (1994) Endothelin releasing stimuli and calcium antagonist in normal and pathological conditions. J Hypertens 12 [Suppl 1]: S27–S31
19. Hassi J, Rintamaki H, Ruskohao H, Leppaluoto J, Vuolteenaho O (1991) Plasma levels of endothelin-1 and atrial natriuretic peptide in men during a 2-hour stay in a cold room. Acta Physiol Scand 142:481–485
20. Koyama H, Tabatu T, Nishizawa Y, Inoue T, Morii H, Yamaji (1989) Plasma endothelin levels in patients with uremia. Lancet 333:991–992
21. Warrens AN, Cassidy MJD, Takahashi K, Ghotei MA, Bloom SR (1990) Endothelin in renal failure. Nephrol Dial Transplant 5:418–422
22. Shichiri M, Hirata Y, Ando K, Kanno K, Emori T, Ohta K (1990) Postural change and volume expansion affect plasma endothelin levels (letter). JAMA 263:661
23. Kaufman H, Oribe E, Oliver JA (1991) Plasma endothelin during upright tilt: relevance for orthostatic hypotension. Lancet 338:1542–1545
24. Emori T, Hirata Y, Ohta K, Kanno K, Eguchi S, Imai T (1991) Cellular mechanism of endothelin-1 release by angiotensin and vasopressin. Hypertension 18:165– 170
25. Giussani M, Cianci M, Gazzano G, Airoldi F, Turolo L, Ghio F, Martinoni A, Zanchetti A, Morganti A (1994) High levels of angiotensin II can increase endothelin-1 production in patients with renovascular hypertension. Proceedings of the 15th meeting of the International Society of Hypertension. Melbourne 1994. J Hypertens 12 [Suppl 3]:S192

The Role of Endothelin in Congestive Heart Failure

A. L. Clavell and J. C. Burnett, Jr.

Introduction

Endothelin (ET) is an endothelium-derived peptide that produces sustained contraction of arterial and venous vascular smooth muscle in vitro [1, 2]. The endothelins which are comprised of three distinct isoforms, ET-1, ET-2, and ET-3 are widely distributed and mediate their biological actions by interacting with at least two receptors which have been cloned and well characterized [3, 4]. The endothelin-A (ET_A) receptor which is widely distributed and preferentially expressed in vascular smooth muscle cells mediates potent vasoconstrictor actions and binds preferentially the ET-1 isoform [5]. The endothelin-B (ET_B) receptor which is preferentially expressed in vascular endothelial cells and binds all three identified endothelin isoforms, ET-1, ET-2, and ET-3, and when activated releases nitric oxide and prostacyclin [5]. Studies have also demonstrated in vitro that ET may enhance smooth muscle and cardiac cell growth supporting a role for ET in cell hypertrophy [6, 7]. In vivo investigations have repeatedly established that pharmacologic concentrations of ET result in potent coronary, renal and systemic vasoconstriction associated with activation of the renin–angiotensin–aldosterone system (RAAS) [8, 9]. Such actions are consistent with a role for ET as an important cardiovascular regulatory peptide of endothelial origin.

While the biological actions of ET at pharmacological concentrations are indisputable, a role for ET in physiologic regulation remains to be established. However, its pathophysiologic importance is suggested by its plasma elevation in disorders of cardiovascular and renal function. Specifically, plasma ET concentrations have been reported to be increased two to three-fold in congestive heart failure (CHF), two-fold in essential hypertension, three-fold in cardiogenic shock and two-fold in transplantation associated hypertension and radiocontrast-induced nephropathy [10–16].

We have recently focused upon its potential role in CHF, a syndrome characterized by low cardiac output, sodium and water retention, and systemic vasoconstriction with activation of the RAAS. Indeed, this syndrome may serve as a functional paradigm in understanding the potential physiologic role for ET as a local paracrine factor as well as a circulating hormone which participates in the maintenance of arterial pressure during chronic reductions of cardiac output. As the syndrome of low-output heart failure progresses, compensatory responses develop that serve to maintain adequate perfusion to vital regional circulations. The heart attempts to compensate by cardiac dilatation and hypertrophy. Extra-cardiac manifestations of CHF include sodium and water retention which increase preload and increase cardiac output via the Frank-Starling mechanism, and activation of vasoconstrictor systems which attempt to maintain adequate perfusion pressure. Vasoconstrictor systems known to be activated include the sympathetic nervous system, the RAAS and vasopressin. In addition, emerging evidence supports a role for these local and circulating humoral systems in the cardiac remodeling which characterizes chronic CHF [17].

Although ET has been considered primarily a local paracrine factor, two to threefold increases in circulating ET have been demonstrated in both experimental CHF produced by rapid ventricular pacing as well as in humans with symptomatic CHF [10–12, 18]. It therefore seems appropriate to consider ET as a circulating hormone in CHF along with other vasoconstrictors activated in heart failure. Given this activation of ET in heart failure, several questions emerge including the mechanisms of its elevation, the functional significance of such plasma elevation in CHF, its role as a prognostic marker in CHF, its interaction with other vasoactive factors known to be activated in CHF and therapeutic strategies designed to antagonize this important peptide hormone and paracrine factor. In this chapter, we will address these questions and speculate as to where future efforts might be directed.

Mechanisms of Increased ET in CHF

Increases in circulating ET have been documented in states of severe cardiovascular stress, including cardiogenic shock, acute myocardial infarction, after major abdominal surgery and after liver transplantation [12, 14, 19, 20]. These increases are consistent with a role for ET as an important endogenous vasoconstrictor modulating vascular tone. Because ET mRNA is widespread[21], increased ET in these pathophysiological states could originate from several sources including heart, lung and peripheral vasculature. Indeed in preliminary studies we have found increased local tissue ET activation in the heart, lung, aorta and kidney in experimental heart failure [22].

Mechanisms for increased plasma ET in CHF include increased production stimulated by either hemodynamic or neurohumoral factors, or decreased clearance and/or metabolism of the peptide. With respect to hemodynamic factors, Margulies et al. demonstrated that right and left atrial pressures were independently and significantly correlated with plasma ET in experimental CHF [11]. These correlations are consistent with a report in which the highest level of plasma ET was observed in acute pulmonary edema, presumably in association with marked increases in left atrial pressure [12]. Because recent observations suggest that cardiac and pulmonary production of ET occurs [21], as it has been observed in preliminary studies where ET-1 mRNA was found to be increased in lung and heart of dogs with experimental heart failure [23], it is tempting to speculate that elevated cardiac filling pressures, and perhaps chronic cardiac or pulmonary vascular distention, are stimuli for ET production or secretion, as has been demonstrated for atrial natriuretic factor (ANF) [24, 25]. Further supporting this concept is the observation by Cody et al. for a strong correlation between circulating ET and pulmonary hypertension in CHF [26]. However, the fact that even higher levels of plasma ET were observed in a model of low CO produced by chronic thoracic inferior vena caval constriction (TIVCC) without atrial distention demonstrates that increased cardiac filling pressures are not essential for the elevation of ET in CHF [27]. Nonetheless, it is possible that increased venous pressure below the inferior vena cava constriction represents an important hemodynamic stimulus for increased ET release. From this perspective, the correlation seen by Margulies et al. was one not related to atrial pressures directly but related to increased peripheral and pulmonary venous pressures [11]. Alternatively, considering the preferential venous vasoconstrictor action of ET in vitro [2], increased cardiac filling pressures may represent an effect rather than a cause of increases in circulating ET. Recent reports suggest that ET release may be modulated by baroreflexes during postural changes, and that postural increases in ET are of hypophyseal origin [28]. However, because CHF is characterized by abnormal responses to baroreceptor unloading, with decreased sympathetic activation [29], the role of neurogenic modulation of ET release in CHF remains speculative.

Elegant studies by Sharefkin et al. have provided insight into an additional mechanism for ET synthesis and release in CHF [30]. In these studies in cultured endothelial cells, an inverse relationship between shear stress and ET synthesis and release was observed. These increases in local ET synthesis and release, in turn, might exert potent mitogenic actions on adjacent vacular smooth muscle in the setting of reduced vascular shear stress. Because CHF is characterized by low cardiac output, hypotension and perhaps decreases in shear stress, the study by Sharefkin provides a possible mechanism for the increase in circulating ET. Such a concept has recently been extended in vivo by Miller and Burnett in a model of chronic increases in blood flow in canine femoral arteries secondary to an arteriovenous fistula. This study demonstrated a decrease in ET content in arteries exposed to the increased flow suggesting that changes in shear stress may play a role in the local regulation of ET production and release [31].

Widespread expression of ET mRNA suggests multiple potential sites as sources of possible excessive local ET production with plasma spill-over [21]. In this regard, Sandok et al. demonstrated a role for the kidney as a site for ET release in the setting of markedly reduced renal perfusion pressure [32]. In this recent investigation, suprarenal aortic constriction markedly increased circulating ET. The relationship between reductions in renal perfusion pressure, as observed in aortic constriction or severe CHF, and increases in circulating ET may be multifactorial. Possible mechanisms include the mechanical stimulus of decreased shear stress and/or the activation of chemical stimuli such as angiotension II and/or hypoxia [30, 33–35]. From this perspective, the kidney may play a key role in the increase in plasma ET during reduced perfusion pressure by the release of a factor that stimulates systemic ET release. In fact, angiotension II (A II) has been shown to stimulate ET production in vitro [36], and the high renin CHF model produced by TIVCC is associated with greater increases in plasma ET than the low renin model produced by rapid ventricular pacing [11, 27]. Supporting such a link between A II and ET production in CHF are preliminary studies which demonstrated that chronic inhibition of angiotensin converting enzyme (ACE) in experimental CHF abolishes the activation of the local tissue ET system [37].

Changes in ET clearance and metabolism might also contribute to increases in circulating ET in CHF. For example, in studies by Cavero et al. [10] low dose exogenous ET produced greater increments in plasma ET in experimental CHF than in normal controls. Moreover, in the TIVCC model, Underwood et al. has demonstrated that whole body metabolic clearance for ET is decreased and may contribute to the striking increases in plasma ET in this setting (unpublished data). In addition, a recent report demonstrated down-regulation of endothelin receptors in cardiac and renal tissue of rabbits with experimental CHF [38]. Furthermore, additional studies have suggested an important role for the ET_B receptor in the clearance of ET-1 [39]. Thus down-regulation of clearance receptors may be another mechanism contributing to the increases in circulating ET in heart failure.

While multiple mechanisms may contribute to the elevation of circulating ET in CHF, its elevation may also have diagnostic implications. In the study of Margulies and co-workers from our laboratory, despite overlap, the calculated sensitivity (85.7%) and specificity (82.1%) of plasma ET as a predictor of CHF were high [11]. Thus, circulating ET may serve as an important marker for CHF. Such a concept is supported by the investigations of Rodeheffer et al. in humans with chronic CHF in which circulating ET increased progressively with the severity of CHF [18]. More recently, plasma ET determination at the time of acute myocardial infarction was a strong and independent predictor of mortality at one year following infarction [40].

Functional Importance of Increased ET in CHF

The functional significance of elevated circulating pathophysiologic concentrations of ET as observed in CHF have been recently addressed in studies by Lerman et al. [41]. In this study, exogenous ET was administered to anesthetized dogs at a concentration to increase plasma ET twofold as observed in CHF. In this study, pathophysiologically relevant increases in circulating ET resulted in significant systemic and renal vasoconstriction in association with a decrease in heart rate and cardiac output in the absence of an increase in arterial pressure. These studies are consistent with an important role for ET in cardiorenal regulation.

However, in pacing-induced CHF, with twofold increases in baseline ET, the systemic and renal vasoconstrictor responses to low dose exogenous ET were attenuated [10]. Possible explanations for such attenuated responses to exogenous ET include receptor downregulation related to sustained increases in ET, the presence of a preconstricted vasculature or the concomitant activation of opposing endogenous vasodilator systems. Employing the model of chronic TIVCC, Underwood et al. observed that the same dose of exogenous ET demonstrated a full vasoconstrictor response despite chronic increases in ET and intense systemic and renal vasoconstriction at baseline [27]. These studies suggest that neither receptor down-regulation nor diminished vasoconstrictor reserve represent dominant mechanisms accounting for reduced hemodynamic responses to exogenous ET in pacing induced CHF.

More recently, preliminary studies in this model of experimental CHF produced by TIVCC strongly suggests that the activation of ET functions to maintain arterial pressure despite decreased cardiac output. In this model of severe renal and systemic vasoconstriction, ET_A receptor blockade resulted in a marked decrease in arterial pressure [42]. Such a response was also observed in a rat model of CHF in response to ET_A and ET_B receptor antagonism [43].

An important distinction between CHF and TIVCC is the presence of atrial stretch and marked increases in ANF in CHF but no atrial distention or increased ANF in the preload limited TIVCC model. To define the role of ANF in modulating the biological responses to exogenous ET, Underwood et al. acutely administered exogenous ANF in the presence of chronic TIVCC. In this setting with pathophysiologic levels of ANF, vascular responses to exogenous ET were markedly attenuated as was observed in pacing induced CHF [10, 27]. This suggests an important role for ANF in counter-regulating the systemic vasoconstrictor actions of ET. In contrast to the systemic circulation, ANF appeared to have little modulating effect on the potent renal actions of ET in the TIVCC model. This may reflect the enhanced sensitivity to ET present in the renal vascular bed, compared with other vascular beds [10] or a dominance of ET actions within this vascular bed [44].

Endothelium-derived relaxing factor (EDRF), presumed to be nitric oxide (NO), is another important endogenous vasodilator [45–47] which may functionally attenuate responses to ET in CHF. Employing the competitive inhibitor of EDRF, NG-monomethyl-L-arginine (L-NMMA), recent studies indicate that EDRF contributes to the control of basal systemic and regional vascular resistances [48]. To assess the functional antagonism between EDRF and ET responses, Lerman et al. observed marked potentiation of the systemic, renal, pulmonary and coronary arterial responses to low dose ET [49]. These studies are consistent with the hypothesis that the endogenous EDRF system serves as a functionally important modulator of the vasoconstrictor actions of ET. Recent investigations also suggest that basal EDRF generation may be preserved [50] or enhanced [51] in the setting of chronic CHF. From this perspective, intact or enhanced EDRF activity, like ANF, may contribute to attenuated ET-mediated vasoconstriction in CHF. Moreover, the balance between these endothelial-derived vasoconstrictor and vasodilator factors may represent an important mechanism of altered vascular tone and responses characteristic of CHF.

From the above discussion, it is clear that ET-1 at pathophysiologic concentrations has biological actions in the control of vascular tone. Recent studies would also suggest that the predominant receptor subtype mediating the systemic and regional vasoconstriction response to ET at pathophysiologic doses is the ET_A receptor subtype [52]. However recent studies have clearly demonstrated that the ET_B receptor can mediate vasoconstriction in certain vasculatures but its role in CHF remains to be established [53]. A cautionary note, however, should be added which is that the elevation of plasma immunoreactive ET in CHF has been recently characterized suggesting a significant increase in Big ET-1 in addition to ET-1 [54]. Although Big ET-1 has less potent biological actions than ET-1 this finding suggests enhanced production of ET-1 as a source of circulating ET. Indeed Big ET-1 may act as a reservoir of ET-1 via its conversion to ET-1 by the putative endothelin converting enzyme (ECE). Further studies elucidating the role of Big ET-1 conversion and the therapeutic potentials of ECE inhibition in CHF need to be addressed.

Future Areas of Research and Therapeutics

Recognizing the elevation of ET in CHF and its known potent vasoconstricting effects, it is interesting to speculate about the pathophysiologic significance of ET in CHF. But first, the physiologic role of ET should be considered. From an evolutionary perspective, it is unlikely that ET evolved as a regulator specifically in response to CHF. It is more likely that this peptide system may participate in the regulation of vascular tone perhaps in response to decreases in cardiac output as may occur with hemorrhage which shares clear similarity to heart failure [34]. It is reasonable from the currently available data to conclude that ET has functionally important local and systemic effects that might be activated at either level. At the local level thrombin, an important mediator released during hemorrhage is known to stimulate ET secretion [55], and one could surmise therefore that ET has the function of maintaining vascular constriction, and thus preventing further hemorrhage, while local repair and clotting mechanisms attempt to repair damaged tissue. At the systemic level, in association with significant decreases in effective arterial volume, there is marked activation of the RAAS. Angiotensin II thus could serve as an obvious and potent stimulus to ET secretion [36]. Therefore, ET might function systemically during hemorrhage to enhance fluid retention via renal vasoconstriction and activation of the RAAS to preserve perfusion pressure. Additionally, the potent effects of venoconstriction would attempt to increase preload, possibly further augmenting cardiac output in this setting.

As heart failure is a low cardiac output state similar to that seen in significant hemorrhage, it seems likely that ET may be elevated in heart failure due to the decrease in effective arterial volume. Similar to the RAAS, the vasoconstricting effects of ET may be beneficial early in CHF, augmenting preload via venoconstriction and increasing systemic vascular resistance to maintain perfusion pressure despite offsetting actions of such modulators as EDRF and ANF. With time however, the chronic effects of ET activation likely become excessive and contribute to a continuing deterioration of cardiac function. This deterioration of cardiac function may also be linked to the remodeling of the heart with associated ventricular hypertrophy and myocyte loss [7, 56]. Studies in vitro have documented an important action of ET upon cardiac myocytes which appear to be mediated by highly expressed and widely distributed cardiac ET receptors [57].

Based upon the hypothesis that it may be beneficial to antagonize the effects of ET in heart failure, various strategies may emerge. Recognizing that a unique converting enzyme may exist which processes Big ET to the mature peptide, an inhibitor of the ECE could serve to inhibit production of ET [58]. Decreasing stimuli to ET secretion would be another possible mechanism to antagonize this peptide system. This could be accomplished by decreasing A II concentrations via converting enzyme inhibition, angiotensin receptor antagonism or renin inhibition. Since the development of several specific endothelin receptor antagonists [59–61] numerous studies have shown acute hemodynamic effects in subarachnoid hemorrhage, renal ischemia and CHF animal models characterized by increased circulating ET [43, 61, 62]. More recently chronic ET_A receptor antagonism ameliorated renal disease progression in a rat model of reduced renal mass [63]. Thus, chronic endothelin antagonism may emerge as an important new strategy in the treatment of CHF. Potentiation of ANF, a known inhibitor via cGMP, either with the ANF peptide or an inhibitor of its degradation via blockade of neutral endopeptidase 24.11 could also be utilized [64]. Most interestingly, use of the newly described endothelial derived ANF like peptide C-type natriuretic peptide (CNP) which activates cGMP and is vasoactive, could also serve therapeutically as a local and potent modulator of ET [65, 66]. In addition to potentiating ANF or ANF-like peptides, nitrates appear to have their action through stimulating the same intracellular mechanism as does EDRF. Thus, some of the currently available treatment strategies likely have some of their beneficial effects by counteracting the actions of ET [67].

In summary, ET is a potent vasoconstrictor peptide which is elevated in CHF and likely contributes to the observed vasoconstriction in this pathophysiologic state. This action serves as a biological paradigm of its physiologic role to preserve arterial pressure in response to an underfilling of effective arterial volume. It is likely that therapeutic regimens aimed at countering the potent vasoconstricting and potentially cardiac remodeling effects of ET will emerge as an important new strategy in the management of CHF.

Acknowledgements. This work was supported by grants from the National Institute of Health (HL-36634, HL-07111 and HL-07269). A. C. was supported by an NIH training grant in cardiovascular disease (HL-07111).

References

1. Yanagisawa M, Kurihara H, Kimura, Tomobe SY, Kombayashi M, Mitsui Y, Yazaki Y, Goto K, Masaki T (1988) A novel potent vasoconstrictor peptide produced by vascular endothelial cells. Nature 332: 411–415
2. Miller VM, Komori K, Burnett JC Jr, Vanhoutte PM (1989) Differential sensitivity to endothelin in canine arteries and veins. Am J Physiol 257: H1127–H1131
3. Arai H, Hori S, Aramori H, Ohkubo H, Nakamishi S (1990) Cloning and expression of a cDNA encoding an endothelin receptor. Nature 348: 732–735
4. Sakurai T, Yanagisawa M, Takuwa Y, Miyasaki H, Kimura S, Goto K, Masaki M (1990) Cloning of a cDNA encoding a non-isopeptide selective subtype of the endothelin receptor. Nature 348: 732–735
5. Masaki T, Kimura S, Yanagisawa M, Goto K (1991) Molecular and cellular mechanism of endothelin regulation: Implications for vascular function. Circulation 84: 1457-1468
6. Neuser D, Knorr A, Stasch J, Kazda S (1990) Mitogenic activity of endothelin-1 and -3 on vascular smooth muscle cells is inhibited by atrial natriuretic peptides. Artery 17 (6): 311–324
7. Ito H, Hirata Y, Hiroe M, Tsujino M, Adachi S, Takamoto T, Nitta M, Taniguchi K, Marumo F (1991) Endothelin-1 induces hypertrophy with enhanced expression of muscle-specific genes in cultured neonatal rat cardiomyocytes. Circ Res 69: 209–215
8. Miller WL, Redfield MM, Burnett JC Jr (1989) Integrated cardiac, renal, and endocrine action of endothelin. J Clin Invest 83: 317–320
9. Goetz KL, Wang BC, Nadwed JB, Zhu JL, Leadley RJ Jr (1988) Cardiovascular, renal, and endocrine responses to intravenous endothelin in conscious dogs. Am J Physiol 255: R1064–R1068
10. Cavero P, Miller WL, Heublein DM, Margulies KB, Burnett JC Jr (1990) Endothelin in experimental congestive heart failure in the anesthetized dog. Am J Physiol 259: F312–F317
11. Margulies KB, Hildebrand FL Jr, Lerman A, Perrella MA, Burnett JC Jr (1990) Increased endothelin in experimental heart failure. Circulation 82: 2226–2230

12. Cernacek P, Stewart DJ (1989) Immunoreactive endothelin in human plasma: marked elevation in patients in cardiogenic shock. Biochem Biophys Res Commun 161:562–567
13. Saito Y, Nakao K, Mukoyama M, Imura H (1990) Increased plasma endothelin level in patients with essential hypertension (letter). N Engl J Med 322:205
14. Lerman A, Click RL, Narr BJ, Wiesner RH, Krom RA, Textor SC, Burnett JC Jr (1991) Elevation of plasma endothelin associated with systemic hypertension in humans following orthotopic liver transplantation. Transplantation 51:646–650
15. Margulies KB, Hildebrand FL, Heublein DM, Burnett JC Jr (1991) Radiocontrast increases plasma and urinary endothelin. J Am Soc Nephrol 2:1041–1045
16. Margulies KB, McKinley LJ, Burnett JC Jr (1992) Endothelin in human and canine radiocontrast-induced nephropathy (abstract). J Vasc Res 29:163A
17. Hirsch AT, Talsness CE, Schunkert M, Martin P, Dzau VJ (1991) Tissue specific activation of cardiac angiotensin converting enzyme in experimental heart failure. Circ Res 69 (2):475–482
18. Rodeheffer R, Lerman A, Heublein DH, Burnett JC Jr (1992) Plasma endothelin concentrations are increased in humans with congestive heart failure. Mayo Clin Proc 67 (8):719–724
19. Miyauchi T, Yanagisawa M, Tomizawa T, Sugishata Y, Suzuki N, Fujino M, Ajisaka P, Goto K, Masaki T (1989) Increased plasma concentrations of endothelin-1 in acute myocardial infarction. Lancet 2:53–54
20. Saito T, Yanagisawa M, Miyauchi T, Suzuki N, Matsumoto H, Fujino M, Masaki T (1989) Endothelin in human circulating blood: effects of major surgical stress (abstract). Jpn J Pharmacol 49:215P
21. Nunez D Jr, Brown MJ, Davenport AP, Neylon CB, Schofeild JP, Wyse RK (1990) Endothelin-1 mRNA is widely expressed in porcine and human tissues. J Clin Invest 85:1537–1541
22. Clavell AL, Nir A, Wright RS, Heublein D, Burnett JC Jr (1994) The endothelin system in thoracic inferíor vena caval constriction. J Am Coll Cardiol 453A [Suppl I]:818–823
23. Wei CM, Clavell AL, Yoshizumi M, Lee ME, Burnett JC Jr (1994) Renal and cardiopulmonary endothelin-1 gene expression in experimental congestive heart failure. JASN 5 (3):596
24. Edwards BS, Zimmerman RS, Schwab TR, Heublein DM, Burnett JC Jr (1988) Atrial stretch, not pressure, is the principal determinant controlling the acute release of atrial natriuretic factor. Circ Res 3:191–195
25. Burnett JC Jr, Kao PC, Hu DC, Heser DW, Heublein DM, Granger JP, Opgenorth TJ, Reeder GS (1986) Atrial natriuretic peptide elevation in congestive heart failure in the human. Science 231: 1145–1147
26. Cody RJ, Haas GJ, Binkley PF, Capers Q, Kelley R (1992) Plasma endothelin correlates with the extent of pulmonary hypertension in patients with chronic congestive heart failure. Circulation 85:504–509
27. Underwood D, Aarhus L, Heublein D, Burnett JC Jr (1992) Endothelin in thoracic inferior vena caval constriction. Am J Physiol 263 (32):H951–H955
28. Kaufmann H, Oribe E, Oliver JA (1991) Plasma endothelin during upright tilt: relevance for orthostatic hypotension. Lancet 338:1542–1545
29. Chen JS, Wang W, Bartholet T, Zucker IH (1991) Analysis of baroreflex control of heart rate in conscious dogs with pacing-induced heart failure. Circulation 83:260–267
30. Sharefkin J, Diamond S, Eskin S, McIntire L, Dieffenbach C (1991) Fluid flow decreases preproendothelin mRNA levels and suppresses endothelin-1 peptide release in cultured human endothelial cells. J Vasc Surg 14:1–9
31. Miller VM, Burnett JC Jr (1992) Modulation of NO and endothelin by chronic increases in blood flow in canine femoral arteries. Am J Physiol 263:H103–H108
32. Sandok EK, Lerman A, Stingo AJ, Perrella MA, Gloviczki P, Burnett JC Jr (1992) Endothelin in a model of acute renal dysfunction: modulating action of atrial natriuretic factor. J Am Soc Nephrol 3:196–202
33. Kourembanas S, Marsden PA, Mcquillan LP, Faller DV (1991) Hypoxia induces endothelin gene expression and secretion in cultured human endothelium. J Clin Invest 88:1054–1057
34. Lloyd MA, Sandberg SM, Miller WL, Edwards BS (1991) Activation of plasma renin activity and aldosterone by endogenous endothelin during hypotensive hemorrhage (abstract). Am J Hypertens 4:4A
35. Nir A, Clavell AL, Heublein D, Aarhus L, Burnett JC Jr (1994) Acute hypoxia and endogenous renal endothelin. J Am Soc Nephrol 4:1920–1924

36. Dohi Y, Hahn A, Boulanger C, Buhler F, Luscher T (1992) Endothelin stimulated by angiotensin II augments contractility of spontaneously hypertensive rat resistance vessels. Hypertension 19: 131–137
37. Clavell AL, Mattingly MT, Nir A, Aarhus LL, Heublein DM, Burnett JC Jr (1994) Angiotensin converting enzyme inhibition modulates circulating tissue endothelin activity in experimental heart failure. Circulation 90 (4); pt 2: I452
38. Löeffler BM, Roux S, Kalina B, Clozel M, Clozel JP (1993) Influence of congestive heart failure on endothelin levels and receptors in rabbits. J Mol Cell Cardiol 25: 407–416
39. Fukuroda T, Fujikawa T, Ozaki S, Ishikawa K, Yano M, Nishikibe M (1994) Clearance of circulating ET-1 by ETB receptor in rats. Biochem Biophys Res Commun 199 (3): 1461–1465
40. Omland T, Lie RT, Aakvaak A, Aarsland T, Dickstein K (1994) Plasma endothelin determination as a prognostic indicator of 1 year mortality after acute myocardial infarction. Circulation 89: 1573–1579
41. Lerman A, Hildebrand FL Jr, Aarhus LL, Burnett JC Jr (1991) Endothelin has biological actions at pathophysiological concentrations. Circulation 83: 1808–1814
42. Clavell AL, Wright RS, Thomas MR, Brandt RR, Opgenorth TJ, Burnett JC Jr (1994) Elevated endogenous endothelin mediates systemic vasoconstriction in experimental chronic congestive failure. J Am Coll Cardiol 172A [Suppl I]: 743–753
43. Teerlink JR, Löeffler BM, Hess P, Maire JP, Clozel M, Clozel JP (1994) Role of endothelin in the maintenance of blood pressure in conscious rats with chronic heart failure. Acute effects of the endothelin receptor antagonist Ro 47-0203 (Bosentan). Circulation 90: 2510–2518
44. Hildebrandt FL Jr, Burnett JC Jr (1990) Endothelin at pathophysiologic concentrations is a potent renal vasoconstrictor resistant to atrial natriuretic factor (abstract). J Am Coll Cardiol 15: 156A
45. Furchgott RF, Zawadski JV (1980) The obligatory role of the endothelial cells in the relaxation of arterial smooth muscle by acetylcholine. Nature 188: 373–376
46. Ignarro LJ, Buga GM, Wood KS, Byrns RE (1987) Endothelium-derived relaxing factor produced and released from artery and vein is nitric oxide. Proc Natl Acad Sci USA 84: 9265–9269
47. Palmer RMJ, Ashton DS, Moncada S (1988) L-Arginine is the physiological precursor for the formation of nitric oxide in endothelium-dependent relaxation. Biochem Biophys Res Commun 153: 1251–1256
48. Perrella MA, Hildebrand FL Jr, Margulies KB, Burnett JC Jr (1991) Endothelium-derived relaxing factor in regulation of basal cardio-pulmonary and renal function. Am J Physiol 261: R323–R328
49. Lerman A, Sandok EK, Hildebrand FL Jr, Burnett JC Jr (1992) Inhibition of endothelium-derived relaxing factor enhances endothelin-mediated vasoconstriction. Circulation 85: 1894–1898
50. Drexler H, Lu W (1992) Endothelial dysfunction of hindquarter resistance vessels in experimental heart failure. Am J Physiol 262 (3): H1640–H1645
51. Hildebrandt FL Jr, Perella MA, Burnett JC Jr (1991) The role of endothelium-derived relaxing factor in experimental congestive heart failure (abstract). J Am Coll Cardiol 17 (2): 281A
52. Clavell AL, Stingo AJ, Margulies KB, Brandt RR, Burnett JC Jr (1995) Role of endothelin receptor subtypes in the in vivo regulation of renal function. Am J Physiol/Renal Electrol 268 (37) (in press)
53. Teerlink JR, Breu V, Sprecher U, Clozel M, Clozel JP (1994) Potent vasoconstriction mediated by endothelin ET_B receptors in canine coronary arteries. Circ Res 74: 105–114
54. Wei CM, Lerman A, Rodeheffer RJ, McGregor CGA, Brandt RR, Wright S, Heublein DM, Kao PC, Edwards WD, Burnett JC (1994) Endothelin in human congestive heart failure. Circulation 89: 1580–1586
55. Schini V, Hendrickson H, Heublein D, Burnett JC Jr, Vanhoutte PM (1989) Thrombin enhances the release of endothelin from cultured porcine aortic endothelial cells. Eur J Pharmacol 165: 333–334
56. Ito H, Hirata Y, Adachi S, Tanaka M, Tsujino M, Koike A, Nogami A, Marumo F, Hiroe M (1993) Endothelin-1 is an autocrine/paracrine factor in the mechanism of angiotensin II-induced hypertrophy in cultured rat cardiomyocytes. J Clin Invest 92: 398–403
57. Wang J, Paik G, Morgan JP (1991) Endothelin-1 enhances myofilament Ca2+ responsiveness in aequorin-loaded ferret myocardium. Circ Res 69: 582–589
58. Ikegawa R, Matsumura Y, Tsukahara Y, Takaoka M, Morimoto S (1990) Phosphoramidon, a metalloproteinase inhibitor, suppresses the secretion of endothelin-1 from cultured endothelial cells by inhibiting a big endothelin-1 converting enzyme. Biochem Biophys Res commun 171: 669–675

59. Fukuroda T, Nishikibe M, Ohara Y, Ihara M, Yano M, Ishikawa K, Fukami T, Ikemoto F (1992) Analysis of responses to endothelins in isolated porcine blood vessels by using a novel endothelin antagonist, BQ-123. Life Sci 50: P107–P114
60. Aramori I, Nirei H, Shoubo M, Sogabe K, Nakamura K, Kojo H, Notsu Y, Ono T, Nakanishi S (1992) Subtype selectivity of a novel endothelin antagonist FR-139 317 for the two endothelin receptors in transfected Chinese hamsters ovary cells. Pharmacology 431: 27–131
61. Clozel M, Breu V, Burri K, Cassal JM, Fischli W, Gray GA, Hirth G, Loeffler BM, Muller M, Neidhart W, Ramuz H (1993) Pathophysiologic role of endothelin revealed by the first orally active endothelin receptor antagonist. Nature 365: 759–761
62. Stingo AJ, Clavell AL, Aarhus LL, Burnett JC Jr (1993) Biological role for the endothelin-A receptor in a model of increased endogenous endothelin. Hypertension 22: 62–66
63. Benigni A, Zoja C, Corna D, Orisio S, Longaretti L, Bertani T, Remuzzi G (1993) A specific endothelin subtype A receptor antagonist protects against injury in renal disease progression. Kidney Int 44: 440–444
64. Margulies K, Perrella M, McKinley L, Burnett JC Jr (1991) Angiotensin inhibition potentiates the renal responses to neutral endopeptidase inhibition in dogs with congestive heart failure. J Clin Invest 88: 1636–1648
65. Stingo A, Clavell A, Aarhus L, Burnett JC Jr (1992) Cardiovascular and renal actions of C-type natriuretic peptide. Am J Physiol 262: H308–H312
66. Nir A, Clavell AL, Aarhus LL, Burnett JC Jr (1993) The modulating effect of C-type natriuretic peptide on the hemodynamic actions of endothelin. Am J Hypertens 6 (5): 74A
67. Miller WL, Cavero PG, Aarhus LL, Heublein DM, Burnett JC Jr (1994) Endothelin-mediated cardiorenal hemodynamics and neuroendocrine effects are attenuated by nitroglycerin in vivo. J Hypertens 12 [Suppl 4]: S43–S49

Endothelin as a Regulator of Vascular Tone in Man: Studies in Upper Limb Resistance and Capacitance Vessels

D. J. Webb, W. G. Haynes, and C. J. Ferro

Introduction

In their seminal paper of 1988, Yanagisawa and colleagues [1] described an endothelium-derived 21 amino acid peptide, endothelin-1, with extremely potent and uniquely sustained vasoconstrictor and pressor actions. Endothelin-1 is one of a family of three isopeptides, and is the predominant isopeptide generated by the vascular endothelium [2]. The other members of the family, endothelin-2 and endothelin-3, are more difficult to detect and are probably less important in the human cardiovasculature. Endothelin-1 is generated from a 38 amino acid precursor, big endothelin-1, through the action of a unique 'endothelin converting enzyme' (ECE), a membrane-bound phosphoramidon-sensitive, thiorphan-insensitive neutral metalloprotease [3–6] which may prove an important target for drug development. One form of this enzyme, ECE-1, has recently been cloned [7]. Endothelin-1 exerts its actions through binding to at least two receptors, both of which have been isolated by in vitro expression of cloned human cDNA [8, 9]. ET_A receptors are highly expressed in vascular smooth muscle cells and appear to be the major receptor subtype causing vasoconstriction in large human arteries [10]. These receptors have an agonist potency: endothelin-1 > endothelin-2 ≫ endothelin-3. ET_B receptors are present on the luminal surface of endothelial cells where they have been clearly shown to mediate release of endothelium-dependent vasodilator substances. However, receptors with the pharmacological characteristics of the ET_B subtype are also present in vascular smooth muscle where they can contribute to vasoconstriction [11–13]. These receptors have an agonist potency: endothelin-1 = endothelin-2 = endothelin-3. The contribution of ET_B receptors to vasoconstriction appears to be somewhat variable, depending markedly on species, vessel type and vessel size [10]. An ET_C receptor, selective for endothelin-3, may also exist but has not yet been cloned in a mammalian species. A number of antagonists selective for the known endothelin receptors have now been described [14] and several are currently undergoing development for clinical use.

On the basis of its actions, endothelin-1 has been postulated to play a role in the pathophysiology of several conditions associated with sustained vasoconstriction, including chronic heart failure, hypertension and renal failure, as well as other conditions associated with vasospasm such as Raynaud's, variant angina and subarachnoid haemorrhage [15, 16]. ECE inhibitors, and the recently described endothelin receptor antagonists, might therefore have potential as vasodilator, antihypertensive and spasmolytic drugs. Here, we have reviewed a series of in vivo experiments in forearm resistance vessels and cutaneous capacitance vessels, using agonists and antagonists of the endothelin system, to examine the relevance of endothelin-1 to cardiovascular physiology and pathophysiology in man. Studies in resistance vessels involved brachial artery infusion of drugs and assessment of forearm blood flow using venous occlusion plethysmography [17]. Studies in capacitance vessels were performed using a displacement technique to measure the size of hand veins infused locally with drugs [17]. All studies involved administration of locally but not systemically active drug doses, so avoiding the influence of infused drugs on other organs, such as the heart, kidney or brain, or the stimulation of neurohormonal reflexes consequent on changes in arterial pressure.

The Actions of Endothelin Agonists

Endothelin-1 is a potent vasoconstrictor of human forearm resistance and cutaneous capacitance vessels in vivo, having a characteristically slow onset of action and a sustained effect in both vessel types [18]. Given systemically, endothelin-1 generates a pressor response [19]. It was initially believed that vasoconstriction to endothelin-1 is wholly ET_A receptor mediated and that stimulation of endothelial ET_B receptors, found only in the endothelium, releases dilator agents, such as nitric oxide and prostacyclin, which serve to modulate the vasoconstriction. In hand veins at least, venoconstriction is modulated by release of vasodilator prostaglandins, but not nitric oxide [20]. Also, in forearm resistance vessels, bolus administration of the relatively ET_B selective agent, endothelin-3, causes substantially greater transient vasodilatation than endothelin-1 [21], suggesting that ET_B receptors can mediate vasodilatation, although the experimental circumstances necessary to elicit this response suggest that it may not be physiological.

Against this background, bolus endothelin-3 causes a substantial degree of slowly progressive vasoconstriction, after initial transient vasodilatation has waned [21]. More importantly, the snake venom peptide sarafotoxin S6c, which is more than 30 000-fold selective for the ET_B receptor, also causes vasoconstriction when given by low dose continuous infusion, implying that ET_B receptors do indeed contribute to endothelin-1 mediated vasoconstriction. However, constriction to the ET_B agonists is usually less than to the non-selective ET_A and ET_B agonist, endothelin-1, and 95% confidence intervals for the results suggest that ET_B receptors account for ~30%–75% of the response to endothelin-1. The similarity of responses to sarafotoxin S6c in forearm resistance vessels and cutaneous capacitance vessels of the hand suggest that ET_B receptors may be of widespread functional importance in human blood vessels. The contrast between these findings and those of studies in vitro [10] may reflect the fact that we examined responses in an intact resistance bed, because ET_B receptor mediated vasoconstriction appears to play a relatively greater role in smaller vessels, particularly those responsible for determining resistance. Interestingly, all of the in vitro studies in which human vessels exhibited little or no ET_B mediated arterial vasoconstriction examined vessels $>400\ \mu m$ in diameter. In addition to the influence of vessel size, there may be regional vascular differences. Local injection of the ET_A antagonist PD 147 953 has been shown to completely prevent vasoconstriction of human skin vessels caused by intradermal injection of endothelin-1, suggesting that vasoconstriction is mainly mediated by ET_A receptors in this microvascular bed [22]. The effects of sarafotoxin S6c, as compared to those of endothelin-1, were less in hand veins than in forearm resistance vessels, despite in vitro evidence in vessels from other species that responses to ET_B agonists are greater in veins than arteries [12, 13]. This may reflect a true species difference, because endothelin-1 is about 8-fold more potent as a venoconstrictor than endothelin-3 in human hand veins [23], also suggesting that ET_A receptors predominate in these vessels.

It appears, at least in man, that there are functionally active ET_A and ET_B receptors on vascular smooth muscle cells causing vasoconstriction, and endothelin-1 would have access to both of these receptor subtypes. These findings have implications for the future development of anti-endothelin therapies, because they suggest that full inhibition of vasoconstriction to endogenously generated endothelin-1 may only be obtained using combined $ET_{A/B}$ endothelin receptor antagonists [24] or, alternatively, by using inhibitors of endothelin generation [25].

Physiology of Endothelin-1 in Human Blood Vessels

In order to examine the role of the endothelin system in the regulation of resistance vessel tone, initial experiments were performed to determine whether ECE activity is present in forearm resistance vessels [25]. Big endothelin-1 alone caused a dose-dependent forearm vasoconstriction. This might be a direct result of big endothelin-1 binding to endothelin receptors, although this is unlikely because big endothelin-1 is about 1000 fold less potent than endothelin-1 in binding to endothelin receptors and in causing constriction of isolated vascular tissue in vitro [26]. When big endothelin-1 was co-infused with a mixed inhibitor of ECE and neutral endopeptidase 24.11, phosphoramidon [3–6], at a dose chosen to achieve local concentrations in the forearm approximately ten-fold higher than the IC_{50} of phosphoramidon for ECE in vivo in animals [4], vasoconstriction to big endothelin-1 was abolished [25]. The blockade of constriction to big endothelin-1 by phosphoramidon is unlikely to have been due to inhibition of receptor binding because, in separate experiments, vasoconstriction to endothelin-1 was unaffected by phosphoramidon [25].

Further confirmation that generation of mature endothelin-1 is responsible for forearm vasoconstriction to big endothelin-1, through the action of a phosphoramidon-sensitive ECE, is derived from measurement of the concentrations of endothelin peptides in blood drawn from the antecubital vein draining the forearm. Plasma immunoreactive endothelin-1, big endothelin-1, and the inactive C-terminal fragment of big endothelin-1 (CTF) that is formed during the cleavage of big endothelin-1 by ECE, were measured and have recently been reported in a preliminary form [27]. In these studies, concentrations of big endothelin-1, endothelin-1 and CTF in venous plasma from the infused , but not the control, arm increased significantly during infusion of big endothelin-1. In the presence of phosphoramidon, plasma endothelin-1 did not increase, consistent with inhibition of big endothelin-1 conversion by ECE. These studies, which used measurements of CTF for the first time in in vivo studies, confirm that a phosphoramidon-sensitive ECE is at least partly responsible for the vasoconstriction to big endothelin-1. Most likely, conversion of big endothelin-1 in the forearm occurs via vascular ECE situated within the forearm blood vessels because circulating blood does not exhibit ECE activity [28].

The approximately tenfold lower potency of big endothelin-1 as a vasoconstrictor than endothelin-1 suggests that local ECE converts ~10% of luminally presented big endothelin-1 to endothelin-1. This may underestimate the true activity of vascular ECE if it is situated remote from the lumen, for example on or within vascular smooth muscle cells [29]. ECE-1, a novel membrane-bound neutral metalloprotease that is expressed abundantly in endothelial cells in vivo, and is structurally related to neutral endopeptidase, has recently been cloned. This enzyme showed only moderate selectivity for endothelin-1 compared with the other endothelin isoforms and it remains to be seen whether other functionally distinct ECE enzymes exist. These studies [7] are consistent with the presence of both intracellular processing of endogenous big endothelin-1 and cell surface conversion of exogenous big endothelin-1. Interestingly, compared with 50%–90% conversion of endogenous big endothelin-1 in transfected cells, the conversion of exogenous big endothelin-1 was much less efficient (~10%) and similar to that found in our studies [25].

Of most interest, when the ECE and neutral endopeptidase inhibitor, phosphoramidon, was infused alone via the brachial artery it caused progressive forearm vasodilatation, reaching a maximum, and highly significant, increase of ~40% (Fig. 1). In contrast, infusion of the same dose of the selective inhibitor of neutral

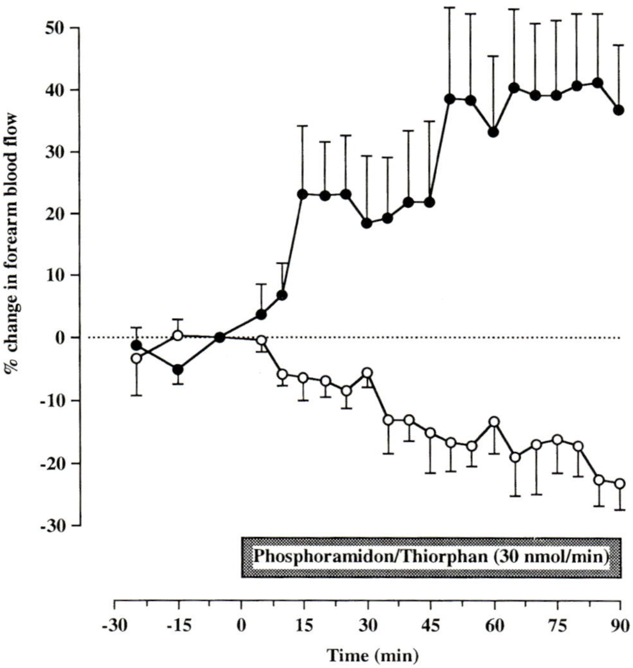

Fig. 1. Progressive forearm vasodilatation to intra-arterial infusion of phosphoramidon (30 nmol/min, ●; $p = 0.03$). In contrast, intra-arterial administration of thiorphan causes forearm vasoconstriction ○; $p = 0.01$; $p = 0.007$ v phosphoramidon). (From [25])

endopeptidase, thiorphan [5, 30], caused forearm vasoconstriction, with a maximum decrease in blood flow of ~20% (Fig. 1). These results suggest that generation of endothelin-1 provides an important contribution to the maintenance of basal vascular tone in man. Although phosphoramidon also inhibits neutral endopeptidase, an enzyme that degrades a number of peptide hormones including the natriuretic peptides, angiotensins and endothelins [31, 32], the forearm vasoconstriction that occurred with thiorphan makes it likely that the forearm vasodilatation is mediated by ECE. Interestingly, ECE-1 is phosphoramidon-sensitive but thiorphan-insensitive [7], suggesting that it was reasonable to use these inhibitors to determine the role of ECE in our studies. The small degree of vasoconstriction caused by thiorphan is probably due to impaired breakdown of constrictor peptides, such as angiotensin II and endothelin-1 [31, 32]. Given the slow onset of vasoconstriction to thiorphan, and the failure of phosphoramidon to inhibit vasoconstriction to endothelin-1, an effect on the metabolism of endothelin-1 is indicated. Such vasoconstriction may explain why systemic inhibition of neutral endopeptidase fails to decrease blood pressure in hypertensive patients and normotensive subjects despite causing natriuresis [33].

Given the potential importance of the findings with phosphoramidon and thiorphan, further experiments were performed using a selective peptide antagonist at the ET_A receptor, BQ-123 (Cyclo[-D-Asp-L-Pro-D-Val-L-Leu-D-Trp-]) [34–36], to investigate the role of endothelin-1 in maintenance of basal resistance vessel tone in man. The dose of BQ-123 (3 mg/h) was chosen because local concentrations achieved in the forearm would be ten-fold higher than the pA_2 of BQ-123 for the ET_A receptor, yet ten-fold lower than those shown to act at the ET_B receptor [34]. In addition, BQ-123 at these or higher concentrations does not affect responses to potassium, noradrenaline, phenylephrine, angiotensin I, angiotensin II, acetylcholine and nitroglycerine in animals [34–36].

First, BQ-123 was shown to abolish vasoconstriction to co-infused endothelin-1. In fact, the net result was slight vasodilatation. Then, when BQ-123 was infused alone, it was found to cause progressive vasodilatation, with a maximum increase of blood flow of ~65% (Fig. 2). These studies, in healthy humans, show for the first time in any species that endogenous generation of endothelin-1 plays a substantial physiological role in the regulation of basal vascular tone. The trend to greater vasodilatation with BQ-123 than with phosphoramidon may be a consequence of inhibition of neutral endopeptidase by the latter. The results with BQ-123 show that endogenous generation of endothelin-1 maintains vascular tone in man through activation of ET_A receptors. An additional role for ET_B receptors cannot presently be excluded.

Interestingly, we have recently had the opportunity to examine the effects of systemic endothelin receptor antagonism in man. Here, we found that maximum doses of a combined $ET_{A/B}$ receptor antagonist, TAK-044, which were well tolerated, caused a small reduction in blood pressure in association with a much more marked reduction of systemic vascular resistance [37], suggesting that the local effects of endothelin antagonism on forearm vascular resistance hold true for the major resistance beds. In addition, TAK-044 appears to be capable of abolishing responses to exogenous endothelin-1 in man, consistent with a role in the management of endothelin-induced vasospasm.

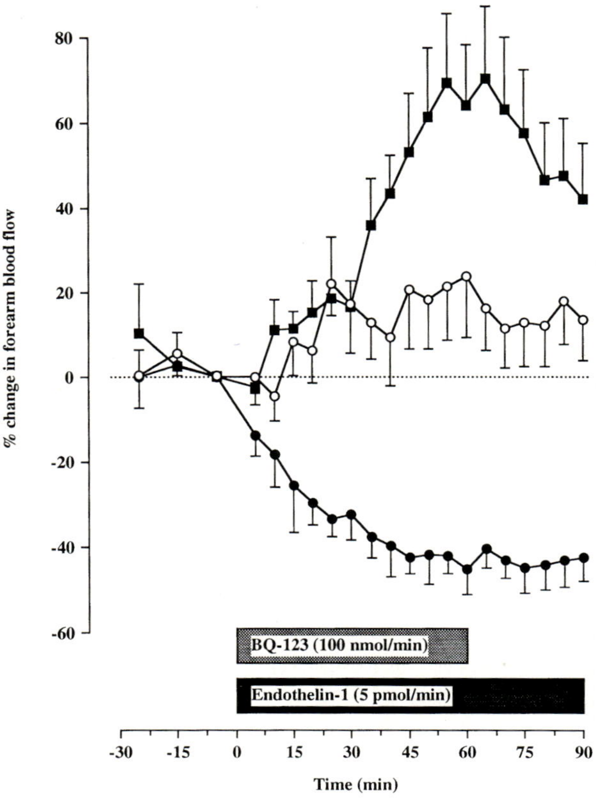

Fig. 2. Abolition of vasoconstriction to intra-arterial infusion of endothelin-1 (5 pmol/min, ●) by co-infusion of BQ-123 (100 nmol/min, ○; $p = 0.006$ v endothelin-1). Infusion of BQ-123 alone causes progressive forearm vasodilatation (■; $p = 0.007$). (From [25])

Pathophysiology of Endothelin-1 in Human Blood Vessels

Hypertension

Established essential hypertension is a condition characterised by a high blood pressure in association with increased peripheral vascular resistance. Cardiac and vascular hypertrophy may also develop. In the early stages, and in borderline hypertension, cardiac output appears to increase and sympathetic tone may be enhanced. The mechanisms causing essential hypertension are still poorly understood. Endothelin-1 is an attractive target for investigation in hypertension because it has potent vasoconstrictor and pressor properties, it is mitogenic and can cause vessel hypertrophy, and it appears to enhance sympathetic function in vitro [38].

Although plasma immunoreactive endothelin concentrations were reported to be elevated in early studies in hypertensive patients, these studies included patients with renal impairment from severe and accelerated phase hypertension, in whom clearance of endothelin-1 is reduced. Studies in well-characterised hypertensive patients with normal renal function have shown similar concentrations of endothelin-1 to those in well-matched normotensive subjects [39, 40]. Indeed, in one of these studies [39], there was a negative correlation between blood pressure and plasma endothelin-1 in the hypertensive group, making a global increase in generation of endothelin-1 unlikely as a contributory factor in essential hypertension.

In studies in rats, conduit and resistance vessels from the spontaneously hypertensive rat (SHR) are more sensitive to the effects of endothelin-1 than those from the Wistar-Kyoto normotensive controls (WKY), and systemic doses of endothelin-1 in vivo have greater pressor effects in SHR than WKY rats [41]. However, in studies designed to examine vascular sensitivity to endothelin-1 in hypertension it is important to be aware that the development of vascular hypertrophy will tend to amplify responses in the hypertensive vessels [42]. Therefore, a recent study in hypertensive subjects designed to investigate whether vasoconstriction to endothelin-1 is enhanced in patients with essential hypertension compared to well-matched normotensive control subjects [40] was performed in hand veins because it is known that these vessels do not develop vascular hypertrophy [43]. Sympathetic responses were also examined in this study becauses endothelin-1 has been shown to potentiate sympathetic responses in vitro [38].

Endothelin-1 caused a slow-onset venoconstriction in both groups of subjects, with the maximum effect reached by 90 min. However, maximal venoconstriction to endothelin-1 was substantially, and very significantly, greater in hypertensive patients than in control subjects (Fig. 3). In contrast, and as shown previously [43], responses to the alpha adrenoceptor agonist, noradrenaline, were not different between

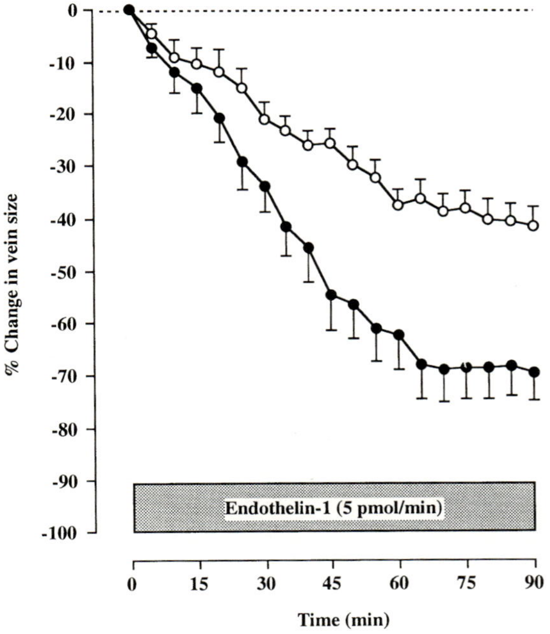

Fig. 3. Venoconstriction to endothelin-1 is significantly greater in hypertensive (●) than in normotensive subjects (○: $p = 0.004$). (From [40])

the two groups. These studies [40] show that patients with essential hypertension have enhanced venoconstriction to endothelin-1. These results might be explained by decreased local venous endothelin-1 generation, leading to endothelin receptor upregulation, although plasma endothelin concentrations were similar to those in normotensive subjects. There was a negative correlation between the degree of venoconstriction to endothelin-1 and blood pressure in the normotensive subjects, suggesting that this phenomenon is not occurring solely as a consequence of the increase in blood pressure (Fig. 4). A causal relationship with the elevation of blood pressure was supported by the positive correlation in the hypertensive subjects (Fig. 4).

In these experiments [40], sympathetically mediated venoconstriction was also markedly enhanced by endothelin-1 in the hypertensive patients but unaffected by endothelin-1 in the control subjects (Fig. 5). Enhanced facilitation of sympathetic vasoconstriction appears to be a separate phenomenon because the lack of basal tone in these vessels precludes an explanation for enhanced venoconstriction to endothelin-1 by potentiated sympathetic responsiveness. In addition, the sympathetic effect appears to be confined to hypertensive patients because it was not seen in a separate study in the forearm resistance vessels of normotensive subjects [44].

Our results are different from previous in vitro work showing an apparently diminished efficacy of endothelin-1 in isolated small arteries of hypertensive patients after correction for media hypertrophy [45]. This difference may be accounted for by methodological differences, such as the in vivo nature of the dorsal hand vein technique, the avoidance of local anaesthesia, which might influence sympathetic responses, and the absence of vessel wall hypertrophy. Alternatively, the difference may be

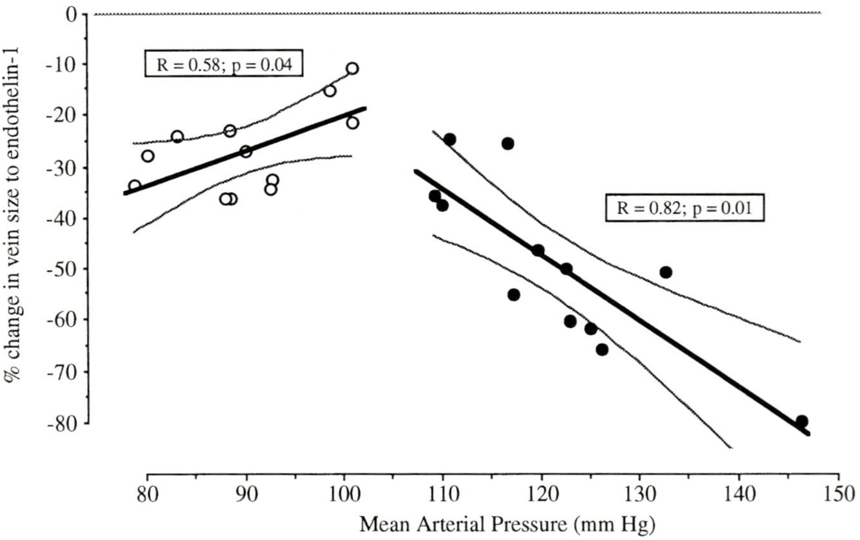

Fig. 4. Regression curves (with 95% confidence intervals) for the correlation between blood pressure and change in vein size to endothelin-1 in normotensive (○) and hypertensive (●) subjects. There is a negative correlation in normotensive subjects but a positive correlation in hypertensive subjects. (From [40])

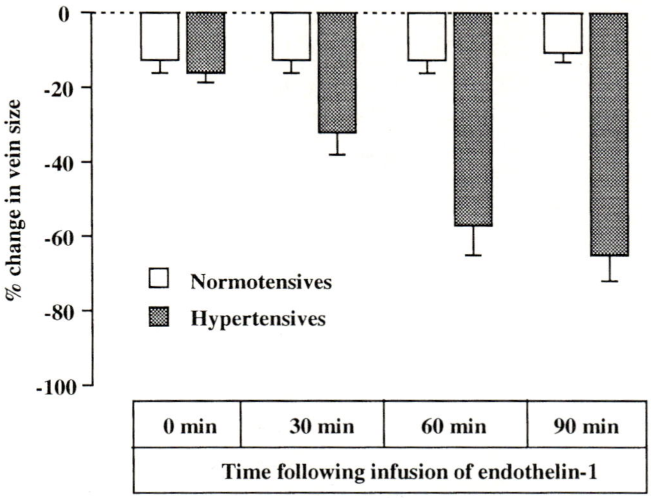

Fig. 5. Sympathetically mediated venoconstriction induced by single deep breath is significantly potentiated in the hypertensive patients but not in the normotensive subjects ($p = 0.001$ vs. normotensives). (From [40])

due to the type of vessel studied. However, in diseases associated with abnormalities in resistance or conduit vessels, such as essential hypertension and Raynaud's disease, similar abnormalities are found in hand veins [46, 47].

There are several potential mechanisms for enhanced venous responsiveness to endothelin-1 in essential hypertension. First, there may be decreased local venous endothelin-1 generation. This is unlikely given that plasma endothelin concentrations were similar between the two groups, though not excluded given that endothelin release is primarily abluminal. Thus, it is possible, though perhaps unlikely, that our findings are explained by increased endothelin receptor number of affinity caused by decreased vascular generation of endothelin-1. Second, the modulating influence of endothelium-derived vasodilator substances on endothelin-1 induced venoconstriction may be abnormal in essential hypertension, resulting in enhanced responses without changes in generation or sensitivity. This would presuppose that impaired endothelial function in hypertension is not the result of elevated pressure per se. There is no published work examining endothelium-dependent responses in veins of hypertensive subjects, although insulin-mediated venodilatation, which may be endothelium-dependent, is clearly abnormal in mild hypertension [46]. However, the unaltered response to noradrenaline, which has endothelial effects, would suggest that endothelial abnormalities are unlikely to fully account for our findings. Third, it is possible that these findings represent an abnormality of venous smooth muscle in hypertension, perhaps affecting receptor subtype expression, receptor number or receptor affinity, or involving the second messenger systems.

These findings in capacitance vessels suggest that exaggerated responsiveness to endothelin-1 may contribute to reduced venous compliance in hypertension. This may, in turn, contribute to the raised cardiac preload and cardiac output observed in the early stages of essential hypertension [48, 49]. Further in vivo studies in resistance vessels will help to show whether there is a global abnormality in endothelin responsiveness in essential hypertension, although such studies will need to take account of structural changes in these vessels. The role of endothelin in hypertension may be further clarified with the use of endothelin receptor antagonists, which are currently entering clinical investigation [16] and it is of major interest that these agents are capable of reducing systemic vascular resistance and arterial pressure [37]. Also, in preliminary arterial studies, we have evidence that the ET_A antagonist, BQ-123, and the ECE inhibitor, phosphoramidon, act as vasodilators of forearm resistance vessels in hypertensive patients [50]. At present, it seems unclear whether drugs based on blocking the endothelin system will be selectively 'anti-hypertensive' rather than hypotensive. However, based on vasodilator, anti-mitogenic and sympatholytic properties, they may offer particular advantages in the treatment of hypertension. Indeed, atherosclerotic human blood vessels show strong immunostaining for endothelin-1, and patients with atherosclerosis have plasma endothelin-1 concentrations that are raised in proportion to the number of affected vessels [51] and overexpression of endothelin-1 mRNA occurs in the vascular smooth muscle of atherosclerotic human vessels [52].

The development of endothelin receptor antagonists and ECE inhibitors for the treatment of hypertension may well prove to be a major focus for clinical investigation. As in other cardiovascular diseases, it remains to be seen whether an ET_A receptor antagonist, leaving endothelial ET_B receptors unopposed but leaving vascular ET_B receptors unblocked, will prove superior to a combined antagonist at both receptors, although the preliminary evidence supports the use of combined antagonists [24, 37]. ET_A antagonists that also inhibit the vascular, but not endothelial, ET_B receptor would, in principle, seem ideal and, interestingly, there is already some evidence that these receptors can be distinguished pharmacologically [53].

Congestive Heart Failure

Chronic heart failure is a common condition with a high morbidity and mortality [54]. It is associated with the development of increased peripheral and renal vascular resistance, associated with abnormal vascular structure and tone. Of recent therapeutic approaches to heart failure, vasodilatation has been the most effective [55, 56]. Hence, the vasoconstrictor, mitogenic and anti-natriuretic properties of endothelin-1 have attracted attention to its potential pathophysiological role in chronic heart failure.

Plasma endothelin concentrations are increased two- to threefold in patients with chronic heart failure [57] and correlate closely with the degree of haemodynamic and functional impairment [58–60]. Big endothelin-1 in particular may be increased, suggesting increased endothelin synthesis rather than reduced clearance, in chronic heart failure [59, 60]. Plasma endothelin concentration also relates to prognosis, with a higher concentration predicting a greater likelihood of death or need for cardiac transplantation [60].

More recently, the ET_A receptor antagonist, BQ-123, and the ECE and neutral endopeptidase inhibitor, phosphoramidon, have been shown to cause substantial peripheral vasodilatation when infused into the forearm circulations of patients with stable chronic heart failure already receiving a loop diuretic and a maximal dose of an angiotensin-1 converting enzyme inhibitor [61]. In contrast, the selective neutral endopeptidase inhibitor, thiorphan, caused vasoconstriction, indicating that ECE inhibition accounted for the vasodilatation caused by phosphoramidon. In comparison to healthy subjects, vasodilatation to BQ-123 tended to be reduced, and to phosphoramidon increased, in patients with heart failure, consistent with up-regulation of ET_B mediated responses. Indeed, more recent studies with endothelin-1 and the ET_B specific agonist, sarafotoxin S6c, are also consistent with this hypothesis [62]. Interestingly, in this regard, circulating concentrations of angiotensin II are increased in heart failure, even after chronic treatment with an ACE inhibitor, and angiotensin II has been shown to down-regulate endothelin-1 binding sites [63] and up-regulate ET_B receptor mRNA [64]. Given that ET_B receptors can mediate venoconstriction [21], it may be that ET_B receptors become of greater functional significance with development of heart failure. Further studies are required to clarify this issue, but these findings again argue for the development of combined $ET_{A/B}$ antagonists. Systemic studies in heart failure must be awaited with interest.

Summary

In studies in the upper limb resistance and capacitance vessels of healthy humans we have shown that both ET_A and ET_B receptors contribute to constriction of vascular smooth muscle, whereas endothelial ET_B receptors modulate vasoconstriction, and can produce transient vasodilatation when stimulated with high doses of ET_B specific agonists [21]. We have also shown for the first time in any species that endogenous generation of endothelin-1 plays a substantial physiological role in the regulation of basal vascular tone, acting at least in part through ET_A receptors [25]. More recent studies confirm that endothelin-1 is responsible for regulation of systemic vascular resistance and blood pressure [37]. In studies in hypertensive patients, we have shown that there is increased responsiveness to endothelin-1 in cutaneous capacitance vessels, and that endothelin-1 selectively increases sympathetic vasoconstriction in the hypertensive subjects [40]. In patients with hypertension and heart failure, endothelin receptor antagonists and ECE inhibitors act as vasodilators. The importance of vasoconstrictor ET_B receptors may be greater in heart failure [61, 62]. The evidence now exists to support the clinical development of agents that inhibit the endothelin system, both in conditions of chronic vasoconstriction, such as hypertension and heart failure, and vasospastic conditions, such as subarachnoid haemorrhage.

Acknowledgement. This work was supported by grants from the Scottish Home and Health Department. CJF is supported by the British Heart Foundation.

References

1. Yanagisawa M, Kurihara H, Kimura S et al (1988) A novel potent vasoconstrictor peptide produced by vascular endothelial cells. Nature 332:411–415
2. Inoue A, Yanagisawa M, Kimura S et al (1989) The human endothelin family: three structurally and pharmacologically distinct isopeptides predicted by three separate genes. Proc Natl Acad Sci USA 86:2863–2867
3. Ikegawa R, Matsumura Y, Tsukahara Y, Takaoka M, Morimoto S (1990) Phosphoramidon, a metalloproteinase inhibitor, suppresses the secretion of endothelin-1 from cultured endothelial cells by inhibiting a big endothelin-1 converting enzyme. Biochem Biophys Res Commun 171:669–675
4. McMahon EG, Palomo MA, Moore WM (1991) Phosphoramidon blocks the pressor activity of big endothelin[1-39] and lowers blood pressure in spontaneously hypertensive rats. J Cardiovasc Pharmacol 17 [Suppl 7]:S29–S33
5. Pollock DM, Opgenorth TJ (1991) Evidence for metalloprotease involvement in the in vivo effects of big endothelin-1. Am J Physiol 261:R257–R263
6. Vemulapalli S, Watkins RW, Brown A, Cook J, Bernardino V, Chiu PJS (1993) Disparate effects of phosphoramidon on blood pressure in SHR and DOCA-salt hypertensive rats. Life Sci 53:783–793
7. Xu D, Enoto N, Giaid A, Slaughter C et al (1994) ECE-1: a membrane-bound metalloprotease that catalyses the proteolytic activation of big endothelin-1. Cell 78:473–485
8. Hosoda K, Nakao K, Arai H et al (1991) Cloning and expression of human endothelin-1 receptor cDNA. FEBS Lett 287:23–26
9. Ogawa Y, Nakao K, Arai H et al (1991) Molecular cloning of a non-isopeptide-selective human endothelin receptor. Biochem Biophys Res Commun 178:248–255
10. Davenport AP, Maguire JJ (1994) Is endothelin-induced vasoconstriction mediated only by ET_A receptors in humans? Trends Pharmacol Sci 15:9–11
11. Seo B, Oemar BS, Siebermann R, Segesser L, Luscher T (1994) Both ET_A and ET_B receptors mediate contraction to endothelin-1 in human blood vessels. Circulation 89:1203–1208
12. Moreland S, McMullen DM, Delaney CL, Lee VG, Hunt JT (1992) Venous smooth muscle contains vasoconstrictor ET_B-like receptors. Biochem Biophys Res Commun 184:100–106
13. Sumner MJ, Cannon TR, Mundin JW, White DG, Watts IS (1992) Endothelin ET_A and ET_B receptors mediate vascular smooth muscle contraction. Br J Pharmacol 107:858–860
14. Haynes WG, Davenport AP, Webb DJ (1993) Endothelin: progress in physiology and pharmacology. Trends Pharmacol Sci 14:225–228
15. Haynes WG, Webb DJ (1993) The endothelin family of peptides: local hormones with diverse roles in health and disease? Clin Sci 84:485–500
16. Webb DJ, Haynes WG (1993) Endothelins come of age. Lancet 342:1439–1440
17. Webb DJ (1995) The pharmacology of human blood vessels in vivo. J Vasc Res (in press)
18. Clarke JG, Benjamin N, Larkin SW, Webb DJ, Davies GJ, Maseri A (1989) Endothelin is a potent long-lasting vasoconstrictor in men. Am J Physiol 257:H2033–H2035
19. Vierhapper H, Wagner O, Nowotny P, Waldhäusl W (1990) Effect of endothelin-1 in man. Circulation 81:1415–1418
20. Haynes WG, Webb DJ (1993) Endothelium dependent modulation of responses to endothelin-1 in human veins. Clin Sci 84:427–433
21. Haynes WG, Strachan FE, Webb DJ (1995) Endothelin ET_A and ET_B receptors cause vasoconstriction of human resistance and capacitance vessels in vivo. Circulation (in press)
22. Wenzel RR, Noll G, Lüscher TF (1994) Endothelin receptor antagonists inhibit endothelin in human skin microcirculation. Hypertension 23:581–586
23. Haefeli WE, Linder L, Kiowski W, Lüscher TF (1993) In vivo properties of endothelin-1 and endothelin-3 in human hand veins and its reversal by bradykinin and verapamil. Hypertension 22:343 (abstract)
24. Clozel M, Breu V, Burri K et al (1993) Pathophysiological role of endothelin revealed by the first orally active endothelin receptor antagonist. Nature 365:759–761
25. Haynes WG, Webb DJ (1994) Contribution of endogenous generation of endothelin-1 to basal vascular tone in man. Lancet 344:852–854

26. Hirata Y, Kanno K, Watanabe TX et al (1990) Receptor binding and vasoconstrictor activity of big endothelin. Eur J Pharmacol 176:225–228
27. Plumpton C, Haynes WG, Webb DJ, Davenport AP (1995) Phosphoramidon inhibits the in vivo conversion of big ET-1 to ET-1 in man. Br J Pharmacol (in press)
28. Watanabe Y, Naruse M, Monzen C, et al (1991) Is big endothelin converted to endothelin-1 in circulating blood? J Cardiovasc Pharmacol 17 [Suppl 7]: S503–S505
29. Ikegawa R, Matsumura Y, Tsukahara Y, Takaoka M, Morimoto S (1991) Phosphoramidon inhibits the generation of endothelin-1 from exogenously applied big endothelin-1 in cultured vascular endothelial cells and smooth muscle cells. FEBS Lett 293:45–48
30. Ollins GM, Krieter PA, Trapani AJ, Spear KL, Bovy PR (1989) Specific inhibitors of endopeptidase 24.11 inhibit the metabolism of atrial natriuretic peptides in vitro and in vivo. Mol Cell Endocrinol 61:201–208
31. Erdos EG, Skidgell RA (1989) Neutral endopeptidase 24.11 (enkephalinase) and related regulators of peptide hormones. FASEB J 3:145–151
32. Abassi ZA, Golomb E, Bridenhaugh R, Keiser HR (1993) Metabolism of endothelin-1 and big endothelin-1 by recombinant neutral endopeptidase EC.3.4.24.11. Br J Pharmacol 109:1024–1028
33. Jardine AG, Connell JMC, Northridge DB et al (1990) The atriopeptidase inhibitor UK 69,578 increases plasma atrial natriuretic factor and causes a natriuresis in normal man. Am J Hypertens 3:661–667
34. Ihara M, Noguchi K, Saeki T et al (1992) Biological profiles of highly potent novel endothelin antagonists selective for the ETA receptor. Life Sci 50:247–255
35. Douglas SA, Elliot JD, Ohlstein EH (1992) Regional vasodilatation to endothelin-1 is mediated by a non-ETA receptor subtype in the anaethetised rat: effect of BQ-123 on systemic haemodynamic responses. Eur J Pharmacol 221:315–324
36. Bazil MK, Lappe RW, Webb RL (1992) Pharmacologic characterization of an endothelin$_A$ (ET$_A$) receptor antagonist in conscious rats. J Cardiovasc Pharmacol 20:940–948
37. Haynes WG, Skwarski KM, Wyld PJ, Ripke H, Webb DJ (1995) Vasodilator effects of the ETA/B antagonist, TAK-044, in man. J Cardiovasc Pharmacol (in press)
38. Yang ZH, Richard V, von Segesser L et al (1990) Threshold concentrations of endothelin-1 potentiate contractions to norepinephrine and serotonin in human arteries: a new mechanism of vasospasm? Circulation 82:188–195
39. Davenport AP, Ashby MJ, Easton P et al (1990) A sensitive radioimmunoassay measuring endothelin-like immunoreactivity in human plasma: comparison of levels in patients with essential hypertension and normotensive control subjects. Clin Sci 78:261–264
40. Haynes WG, Hand MF, Johnstone HA, Padfield PL, Webb DJ (1994) Direct and sympathetically mediated venoconstriction in essential hypertension: enhanced responses to endothelin. J Clin Invest 94:1359–1364
41. Miyauchi T, Ishikawa T, Tomobe Y et al (1989) Characteristics of pressor response to endothelin in spontaneously hypertensive and Wistar-Kyoto rats. Hypertension 14:427–434
42. Folkow B (1978) Cardiovascular structural adaptation: its role in the initiation and maintenance of primary hypertension. Clin Sci Mol Med 55:3S–22S
43. Eichler H, Ford GA, Blaschke TF, Swislocki A, Hoffman BB (1989) Responsiveness of superficial hand veins to phenylephrine in essential hypertension: alpha adrenergic blockade during prazosin therapy. J Clin Invest 83:108–112
44. Cockcroft JR, Clarke JG, Webb DJ (1991) The effect of intra-arterial endothelin on resting blood flow and sympathetically mediated vasoconstriction in the forearm of man. Br J Clin Pharmacol 31:521–524
45. Schiffrin EL, Deng LY, Larochelle P (1992) Blunted effects of endothelin upon small subcutaneous resistance arteries of mild essential hypertensive patients. J Hypertens 10:437–444
46. Feldman RD, Bierbrier GS (1993) Insulin-mediated vasodilatation: impairment with increased blood pressure and body mass. Lancet 342:707–709
47. Bedarida GV, Kim D, Blaschke TF, Hoffman BB (1993) Venodilatation in Raynaud's disease. Lancet 342:1451–1454
48. Ellis C, Julius S (1973) Role of central blood volume in hyperkinetic borderline hypertension. Br Heart J 35:450–455

49. Safar M, London G, Levenson J, Simon A, Chau N (1979) Rapid dextran infusion in essential hypertension. Hypertension 1:615–623
50. Haynes WG, Webb DJ (1994) Generation of endothelin contributes to vascular tone in man. Hypertension 4:380 (abstract)
51. Lerman A, Edwards BS, Hallett JW, Heublein DM, Sandberg SM, Burnett JC (1991) Circulating and tissue endothelin immunoreactivity in advanced atherosclerosis. N Engl J Med 325:997–1001
52. Winkles JA, Alberts GF, Brogi E, Libby P (1993) Endothelin-1 binding and endothelin receptor mRNA expression in normal and atherosclerotic human arteries. Biochem Biophys Res Commun 191:1081–1088
53. Gray GA (1995) Pharmacological characterization of endothelin receptors in the cardiovascular system. In: Gray GA, Webb DJ (eds) Molecular biology and pharmacology of the endothelins. Landes, Austin, Texas, USA (Molecular biology intelligence unit monograph series)
54. Dargie HJ, McMurray JJV (1994) Diagnosis and management of heart failure. Br Med J 308: 321–328
55. The SOLVD Investigators (1991) Effect of enalapril on survival in patients with reduced left ventricular ejection fractions and congestive heart failure. N Engl J Med 325:293–302
56. Cohn JN, Johnson G, Ziesche S et al (1991) A comparison of enalapril with hydralazine-isosorbide dinitrate in the treatment of chronic congestive heart failure. N Engl J Med 325:303–310
57. McMurray JJ, Ray SG, Abdullah I, Dargie HD, Morton JJ (1992) Plasma endothelin in chronic heart failure. Circulation 85:1374–1379
58. Cody, RJ, Haas GJ, Binkley PF, Capers Q, Kelly R (1992) Plasma endothelin correlates with the extent of pulmonary hypertension in patients with chronic congestive heart failure. Circulation 85:504–509
59. Pacher R, Bergler-Klein J, Globits S et al (1993) Plasma big endothelin–1 concentrations in congestive heart failure patients with or without systemic hypertension. Am J Cardiol 71:1293–1299
60. Wei C-M, Lerman A, Rodeheffer RJ et al (1994) Endothelin in human congestive heart failure. Circulation 89:1580–1586
61. Love MP, Haynes WG, Webb DJ, McMurray JJV (1994) Anti-endothelin therapy is of potential benefit in heart failure. Circulation 90:I-547
62. Love MP, Haynes WG, Webb DJ, McMurray JJV (1995) Endothelin in chronic heart failure. J Cardiovasc Pharmacol (in press)
63. Roubert P, Gillard V, Plas P, Guillon JM, Chabrier PE, Braquet P (1989) Angiotensin II and phorbol-esters potently down-regulate endothelin (ET-1) binding sites in vascular smooth muscle cells. Biochem Biophys Res Commun 164:809–815
64. Kanno K, Hirata Y, Tsujino M et al (1993) Up-regulation of ET_B receptor subtype mRNA by angiotensin II in rat cardiomyocytes. Biochem Biophys Res Commun 194:1282–1287

Systemic and Endocrine Effects of Endothelin in Humans

H. Vierhapper

Introduction

The endothelium is the key to a complex system of neural and humoral factors modulating vascular smooth muscle tone by producing both relaxing and constrictive factors including the three currently known members of the endothelin family, endothelin-1 (ET-1 [1]), ET-2 and ET-3. These three structurally related peptides correspond to two types of receptors, desiĝnated ET_A and ET_B. The hemodynamic effects of endothelin-1 have been extensively studied in various animal models. Furthermore an impressive bulk of data has been obtained in vitro and in vivo concerning the recently recognized importance of endothelins as regulatory factors in various endocrine tissues. Although endothelin-1 is secreted mainly abluminally [2] and may play a paracrine rather than an endocrine role, experiments evaluating the effects of pharmacological doses in man may help to interpret the effect of increased plasma concentrations of ET-1 which have been observed in various clinical conditions [3]. However, due to the difficulty arising from the use of such a potent vasopressor agent on an experimental basis in humans comparatively few such experiments have been performed in man.

Blood Pressure

In healthy humans the systemic infusion of 2.0 pmol (5.0 ng) ET-1 kg^{-1} min^{-1} induces a rise in mean blood pressure of about 5 mm Hg [4]. This dose of ET-1 also induces sodium retention and an increase in renal vascular resistance [5]. The absolute reabsorption of sodium and water in the proximal tubulus remains unchanged but in the distal tubules absolute reabsorption of both sodium and water decreases [5, 6]. At considerably higher infusion rates a rise in plasma concentrations of renin, aldosterone and ANP has been induced by ET-1 in the conscious dog [7] whereas doses of ET-1 admissible in man do not alter plasma concentrations of renin, aldosterone, atrial natriuretic peptide, arginine vasopressin and angiotensin II [5–7] but induce a rise in serum potassium concentrations both in man and in the dog [5, 8, 9] in a yet unexplained fashion (Table 1).

Local infusions of ET-1 (5.0 pmol min^{-1} into dorsal hand veins) cause prolonged local venoconstriction but do not induce changes in systolic or diastolic blood pressure [10–12]. Local venoconstriction is potentiated by aspirin [10] suggesting an important role of locally produced vasodilator prostaglandins. In contrast, inhibition

Table 1. Mean blood pressure, pulse rate and serum (plasma) concentrations of potassium (K^+), sodium (Na^+), renin (PRC), aldosterone (PA) and atrial natriuretic factor (ANF) in six healthy humans before, during and after the i.v. administration of endothelin (1.0, 2.5 and 5.0 ng kg^{-1} min^{-1}) (from [4])

Time (min)	BP_m (mm Hg)	PR (× min E-1)	K (mmol/l)	Na (mmol/l)	PRC (GU × 10 E-4/ml)	PA (pmol/l)	ANF (pmol/l)
0	87.1 ± 7.3	66.3 ± 1.6	4.0 ± 0.3	142.7 ± 1.0	0.60 ± 0.20	213.6 ± 27.7	7.0 ± 1.7
+ 15	88.7 ± 8.0	61.3 ± 4.8			0.59 ± 0.21	199.7 ± 38.8	7.1 ± 1.6
+ 30	91.2 ± 7.5***	62.6 ± 3.3			0.61 ± 0.23	197.0 ± 80.4	6.0 ± 2.7
+ 45	92.6 ± 9.2***	58.0 ± 4.2**			0.51 ± 0.22	213.6 ± 80.4	6.8 ± 2.5
+ 60	91.2 ± 6.3***	58.0 ± 4.2**			0.48 ± 0.17	246.9 ± 97.1	6.4 ± 3.2
+ 75	88.3 ± 5.8	61.3 ± 3.3			0.56 ± 0.24	235.8 ± 108.2	6.3 ± 2.2
+ 90	87.8 ± 5.6	58.7 ± 3.3			0.48 ± 0.19	216.4 ± 88.8	5.0 ± 2.0*
+ 105	87.0 ± 6.4	62.0 ± 4.9	4.6 ± 0.2***	139.5 ± 2.3*	0.46 ± 0.18*	235.8 ± 99.9	6.0 ± 2.1

Values for BP_m and PR represent the mean of measurements taken in 5 min intervals during the respective period of time. * $p < 0.05$, ** $p < 0.01$, *** $p < 0.005$ vs respective basal values of endogenous prostaglandin synthesis by indomethacin is of little if any importance for the overall pressor action of ET-1 in healthy humans [8].

During brachial artery infusions of up to 20 pmol (50 ng) ET-1 min^{-1} per 100 ml forearm tissue endothelin-1 exerts a dual action consisting of a transient initial vasodilation followed by a prolonged vasoconstriction [13, 14]. The vasodilator effect is diminished and the vasoconstriction is enhanced in smokers [14]. The initial depressor effect is attributed to the release of prostacyclin or of EDRF [13]. A substantially higher dose (5 nmol min^{-1}) has been infused intra-arterially in one human subject and has resulted in pronounced sideeffects [15].

Plasma concentrations of endothelin during the systemic administration of 2.0 pmol (5.0 ng) ET-1 were above 50 pmol l^{-1} and reached values of more than 100 pmol l^{-1} in single individuals (Fig. 1 [4]). Except for a transient, local erythema no side effects were seen during this initial study. However during a later experiment [16] a prolonged, hypertensive reaction was observed in one individual (Fig. 2). The rise in blood pressure in this individual continued even after the infusion of ET-1 had been stopped. This emphasizes that the biological activity of endothelin-1 is not congruent with its removal from the circulation, which takes place with a half-life of approximately 3.6 min [4]. Blood pressure returned to normal in this volunteer following the oral administration of the calcium channel blocker nifedipine.

Both ET-1 induced sodium retention and the rise in renal vascular resistance can be prevented by pretreatment with nifedipine [5]. Nifedipine and verapamil also reverse endothelin-1 induced vasoconstriction in the human forearm [13]. However, under normal circumstances the pressor effect induced by moderate doses of ET-1 in healthy humans is not altered by pretreatment with 10 mg nifedipine p.o. [8]. Remarkably and most likely due to an altered metabolic clearance rate, induced plasma concentrations of endothelin-1 are higher following pretreatment with nifedipine as compared to control experiments. The seemingly unchanged pressor response to ET-1 in the presence of nifedipine thus indicates at least a relative impact of nifedipine on the action of endothelin-1 [8].

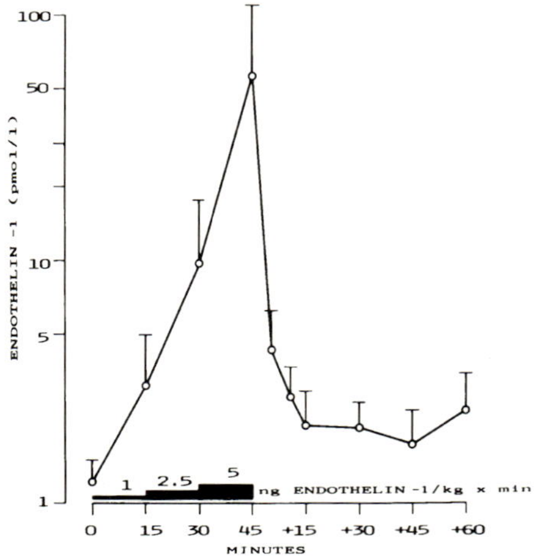

Fig. 1. Plasma concentrations of endothelin-1 (pmol/l) in six healthy humans before, during and after the intravenous administration of endothelin-1 (1.0, 2.5, and 5.0 ng kg^{-1} min^{-1}). (From [4])

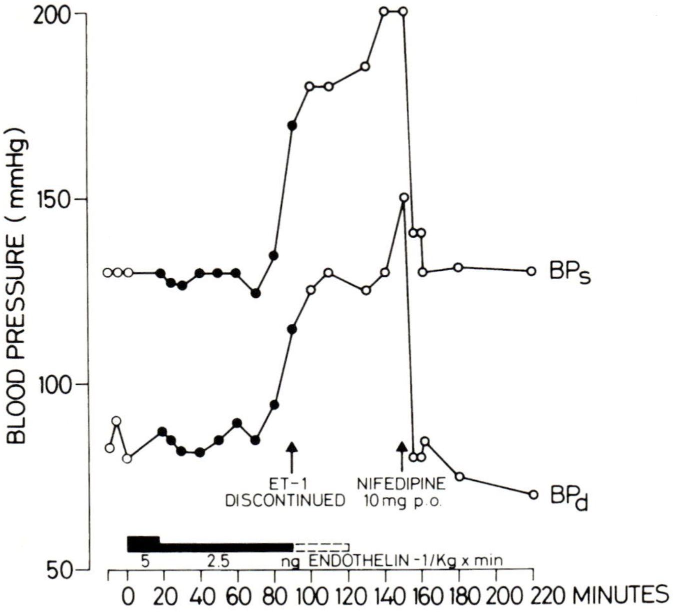

Fig. 2. Endothelin-1-induced rise in systolic (BP_s) and diastolic (Bp_d) blood pressure in a healthy individual. Reversal by nifedipine (10 mg p.o.). (Data from [16])

Pretreatment with cyclosporin A in healthy subjects does not modify the pressor response to ET-1 [8] although this does not argue against the possibility that this drug might have an impact on the action of ET-1 during prolonged treatment, since treatment with cyclosporine increases the density of binding sites for ET-1 in cardiac cell membranes [17].

Cardiac Output and Regional Clearance Rates

In healthy male volunteers even a modest increase in endothelin plasma concentrations achieved by the infusion of 0.4 pmol endothelin-1 kg^{-1} min^{-1} induces a slight decrease in cardiac output and in heart rate [18, 19]. Total systemic vascular resistance increases by about 25% [18]. Splanchnic blood flow decreases in parallel to the decrease in cardiac output [18] while renal perfusion is reduced more substantially, even at such a comparatively low rate of infusion [5, 20]. Splanchnic and renal vascular resistance are increased by 20% and by 50%, respectively, but vascular resistance of the lung and the leg is unchanged [18]. Thus the renal circulation appears to be more sensitive to the effects of endothelin-1 than other areas, a fact that needs to be considered when the administration of larger doses of ET-1 is considered. Sixty minutes after termination of endothelin-1 administration hemodynamic changes tend to return to baseline.

In man the lung is the main site of removal of ET-1 [18] and pronounced extraction during passage through the pulmonary circulation appears to account to a large extent for the short apparent half-life of endothelin-1 in man. This substantial, if incomplete, pulmonary uptake is responsible for the arteriovenous concentration difference of ET-1 in man, the venous concentration being higher than the arterial one [21]. The cerebral circulation apparently does not extract ET-1 [19]. Since during an i.v. infusion of ET-1 the splanchnic, renal and peripheral vascular beds receive blood previously cleared by the lungs their contribution to the overall removel of ET-1 from the circulation is only about 3%, 9% and 6%, respectively, of the infusion rate, although the efficacy of splanchnic ET-1 clearance is comparable to that of the lung. As opposed to the situation in healthy humans a net pulmonary release of ET-1 into the circulation is found in patients with adult respiratory distress syndrome where this defect may well induce a vicious cycle promoting microvascular damage [22].

Adrenal

The potential physiological importance of ET-1 is not limited to its hemodynamic properties. Binding sites for ET-1 have been demonstrated in various endocrine and non-endocrine tissues including the adrenal gland [23]. At least in vitro, ET-1 interferes with adrenal steroid hormone synthesis, particularly the adrenocorticotropin

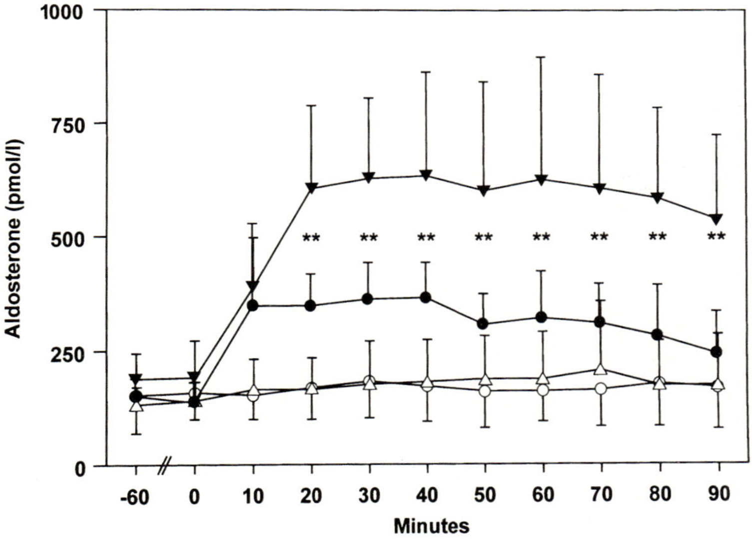

Fig. 3. Effect of endothelin-1 on basal and ACTH-stimulated plasma concentrations of aldosterone (pmol/l) in healthy humans ($n=6$, means ± SD). △–△, Experiment 1: control experiment; ○–○, experiment 2: infusion of endothelin-1; ●–●, experiment 3: ACTH stimulation test (250 μg i.v. in the absence of endothelin-1); ▼–▼, experiment 4: ACTH stimulation test (250 μg i.v. in the presence of endothelin-1). *Asterisks* indicate differences between experiments 3 and 4 at single points of observation (** $p<0.01$). (From [25])

(ACTH)-stimulated release of aldosterone [24]. ACTH-stimulated secretion of cortisol from fasciculata cells remains unchanged and this has led to the suggestion of a specific action of ET-1 on glomerulosa cells, possibly distal to the binding of ACTH to its receptor [24]. As mentioned above, i.v. ET-1 does not influence basal plasma aldosterone concentrations in healthy man, at least in the sodium-replete state [4]. However, this observation does not apply to ACTH-stimulated concentrations of aldosterone. When healthy men receive a systemic infusion of ET-1 (2 pmol [5 ng] kg^{-1} min^{-1} for 15 min followed by 1 pmol [2.5 ng] kg^{-1} min^{-1} for 105 min) together with an i.v. bolus-dose (250 μg) of ACTH the ACTH-induced rise in plasma aldosterone is markedly augmented as compared to a placebo experiment ([25] Fig. 3). Basal and ACTH-stimulated concentrations of cortisol, corticosterone and 18-OH-corticosterone are unchanged by the concomitant infusion of endothelin-1 and this again indicates a specific effect on aldosterone synthesis distal to the initial steps of adrenal steroidogenesis, since exogenous endothelin-1 apperently influences adrenal function in healthy men by selectively augmenting ACTH-induced secretion of aldosterone. Although the endothelin-induced increase in serum potassium does not result in a rise in (basal) plasma aldosterone concentrations [4] it could contribute to the enhanced aldosterone reponse to ACTH. It should be remembered at this point that due to the its unipolar mode of secretion [2] plasma concentrations of endothelin-1 may not necessarily correspond to local (i.e. intra-adrenal) concentrations. In condition such as cardiogenic shock, where plasma concentrations of endothelin-1 are elevated, local concentrations of endothelin-1 may reach levels as induced during infusion studies. Under these circumstances the existence of ET-1 as an independent factor directly enhancing aldosterone secretion could contribute to the maintainance of blood pressure.

Pituitary

The abundant concentration of high-affinity binding sites for endothelins in the human pituitary is inversely related to the peptides' low local tissue concentrations [26], and the importance of endothelins in the regulation of pituitary function in vitro and in animal models is well established [27–30]. Although the presence of endothelin-1 in human cerebrospinal fluid has been known for some time [31, 32] the actual concentrations of different endothelins around the pituitary cells in vivo is not known [33] and it is therefore difficult to evaluate the potential role of endothelins on pituitary hormone secretion in man. In healthy male human volunteers i.v. endothelin-1 (5.0 ng [2.0 pmol] kg^{-1} min^{-1} for 15 min followed by 2.5 ng [1.0 pmol] kg^{-1} min^{-1} for 105 min) exerts a suppressive action on stimulated, though not on basal, plasma concentrations of PRL and GH, but stimulates the release of corticotropin and, possibly, of LH and FSH (Fig. 4 [16]). Analogous experiments performed with and/without pretreatement with nifedipine indicate that these effects are not, or at least not exclusively, mediated by Ca^{2+} channel activation [34]. This is of note, since dihydropyridine-sensitive L-type calcium channels have been identified as as mediators of the metabolic response (i.e. increased rate of glucose utilization) of pars intermedia and distalis to centrally administered endothelin in rats [35]. As pointed out

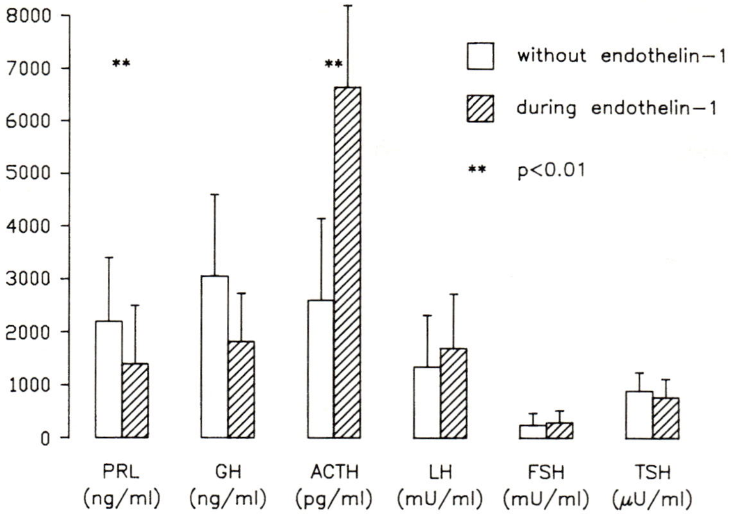

Fig. 4. Effect of endothelin-1 (5 ng/kg^{-1} min^{-1} for 15 min followed by 2.5 ng/kg^{-1} min^{-1} for 105 min) on stimulated pituitary hormone secretion in healthy men ($n=5$; areas under the curve/90 min; means ± SD). Pituitary stimulation was done using luteinizing hormone-releasing hormone (LHRH; 100 μg i.v.); thyrotrophin-releasing hormone (TRH; 400 μg i.v.); corticotrophin-releasing hormone (CRF; 100 μg i.v.); and growth hormone-releasing hormone (GHRH; 100 μg i.v.). *PRL*, prolactin; *GH*, growth hormone; *ACTH*, adrenocorticotrophic hormone; *LH*, luteinizing hormone; *FSH*, follicle-stimulating hormone; *TSH*, thyroid-stimulating hormone. (From [16])

above, plasma concentrations achieved during this dose of systemic endothelin-1 are in the same range (5–10 pmol/l) as those observed in various pathological conditions. Even assuming that haemodynamic effects of endothelin-1 (e.g. by modulating the blood flow through the brain or the pituitary gland) might cause a non-specific activation of the hypothalamic–pituitary axis [36] the obtained results would still be of pathophysiological relevance.

Summary

In pharmacologic doses endothelin-1 exerts both haemodynamic and non-haemodynamic effects in healthy humans including alterations of adrenal and anterior pituitary function. In a variety of pathological conditions an increased secretion and/or a reduced clearance of endothelin-1 results in plasma concentrations comparable to those induced by the peptide's systemic administration. At present it is not clear whether this represents simply a potentially harmful sequel to endothelial injury or whether a rise in plasma ET-1 concentrations may, under certain circumstances, be part of the organism's life-maintaining adaptive mechanisms.

References

1 Yanagisawa M, Kurihara H, Kimura S, Tomobe Y, Kobayashi M, Mitsui Y, Yazaki Y, Goto K, Masaki T (1988) A novel potent vasoconstrictor peptide produced by vascular endothelial cells. Nature 332:411–415
2. Wagner OF, Christ G, Wojta J, Vierhapper H, Parzer S, Nowotny PJ, Schneider B, Waldhäusl W, Binder BR (1992) Polar Secretion of Endothelin-1 by Cultured Endothelial Cells. J Biol Chem 267:16 066–16 068
3. Hutchison FN (1989) Endothelin: a new role for an old friend? J Lab Clin Med 122:126–127
4. Vierhapper H, Wagner O, Nowotny P, Waldhäusl W (1990) Effect of endothelin-1 in man. Circulation 81:1415–1418
5. Rabelink TJ, Kaasjager KAH, Boer P, Stroes EG, Braam B, Koomans HA (1994) Effects of Endothelin-1 on renal function in humans: implications for physiology and pathophysiology. Kidney Int 46:376–381
6. Sorensen SS, Madsen JK, Pedersen EB (1994) Systemic and renal effect of intravenous infusion of endothelin-1 in healthy human volunteers. Am J Physiol 266:F411–F418
7. Goetz KL, Wang BC, Madwed JB, Zhu JL, Leadley RJ (1988) Cardiovascular, renal, and endocrine responses to intravenous endothelin in conscious dogs. Am J Physiol 255:R1064–R1068
8. Vierhapper H, Wagner OF, Nowotny P, Waldhäusl W (1992) Effect of endothelin-1 in man: pretreatment with nifedipine, with indomethacin and with cyclosporine A. Eur J Clin Invest 22:55–59
9. Donckier JE, Hanet C, Berbinschi A, Galanti L, Robert A, Van Mechelen H, Pouleur H, Ketelslegers J-M (1991) Cardiovascular and endocrine effects of endothelin-1 at pathophysiological and pharmacologica plasma concentrations in conscious dogs. Circulation 84:2476–2484
10. Webb DJ, Haynes WG (1993) Venoconstriction to endothelin-1 in humans is attenuated by local generation of prostacyclin but not nitric oxide. J Cardiovasc Pharmacol 22:S317–S320
11. Haynes WG, Webb DJ (1993) Venoconstriction to endothelin-1 in humans: role of calcium and potassium channels. Am J Phyciol 265:H1676–H1681
12. Blöchl-Daum B, Vierhapper H, Eichler H-G, Waldhäusl W (1992) Endothelin-induced venoconstriction is unaffected by type 2-diabetes: in vivo effect of histamine on the endothelin action on veins. Arch Int Pharmacodyn 316:90–96
13. Kiowski W, Lüscher TF, Linder L, Bühler FR (1991) Endothelin-1-induced vasoconstriction in humans. Reversal by calcium channel blockade but not by nitrovasodilators or endothelium-derived relaxing factor. Circulation 83:469–475
14. Kiowski W, Linder L, Stoschitzky K, Pfisterer M, Burckhardt D, Burkart F, Bühler FR (1994) Diminished vascular response to inhibition of endothelium-derived nitric oxide and enhanced vasoconstriction to exogenously administered endothelin-1 in clinically healthy smokers. Circulation 90:27–34
15. Dahlöf B, Gustafsson D, Hedner T, Jern S, Hansson L (1990) Regional haemodynamic effects of endothelin-1 in rat and man: unexpected adverse reaction. J Hypertens 8:811–817
16. Vierhapper H, Hollenstein U, Roden M, Nowotny P (1993) Effect of endothelin-1 in man – impact on basal and stimulated concentrations of LH, FSH, TSH, GH, ACTH and PRL. Metabolism 42: 902–906
17. Nayler WG, Gu XH, Casley DJ, Panagitopoulos S, Liu J, Mottram PL (1990) Cyclosporine increases endothelin-1 binding site density in cardiac cell membranes. Biochem Biophys Res Commun 163: 1270–1274
18. Wagner OF, Vierhapper H, Gasic S, Nowotny P, Waldhäusl W (1992) Regional effects and clearance of endothelin-1 across pulmonary and splanchnic circulation. Eur J Clin Invest 22:277–282
19. Weitzberg E, Ahlborg G, Lundberg JM (1993) Differences in vascular effects and removal of endothelin-1 in human lung, brain, and skeletal muscle. Clin Physiol 13:653–662
20. Gasic S, Wagner OF, Vierhapper H, Nowotny P, Waldhäusl W (1992) Regional hemodynamic effects and clearance of endothelin-1 in humans: renal and peripheral tissues may contribute to the overall disposal of the peptide. J Cardiovasc Pharmacol 19:176–180
21. Wagner OF, Nowotny P, Vierhapper H, Waldhäusl W (1990) Determination of endothelin-1 in human plasma: arterio-venous differences and release during venous stasis. Eur J Clin Invest 20: 502–505

22. Druml W, Steltzer H, Waldhäusl W, Lenz K, Hammerle A, Vierhapper H, Gasic S, Wagner OF (1993) Endothelin-1 in adult respiratory distress syndrome. Am Rev Respir Dis 148:1169–1173
23. Cozza EN, Gomez-Sanchez CE, Foecking MF, Chiou S (1989) Endothelin binding to cultured adrenal zona glomerulosa cells and stimulation of aldosterone secretion. J Clin Invest 84:1032–1035
24. Rosolowsky LJ, Campbell WB (1990) Endothelin enhances adrenocorticotropin-stimulated aldosterone release from cultured bovine adrenal cells. Endocrinology 126:1860–1866
25. Vierhapper H, Nowotny P, Waldhäusl W (1995) Effect of endothelin-1 in man – impact on basal and ACTH-stimulated concentrations of aldosterone. J Clin Endocrin Metab 80:948–951
26. Takahashi K, Ghatei MA, Jones PM, Murphy JK, Lam H-C, O'Halloran DJ, Bloom SR (1991) Endothelin in human brain and pituitary gland. Presence of immunoreactive endothelin, endothelin messenger ribonucleic acid, and endothelin receptors. J Clin Endocrinol Metab 72:693–699
27. Samson WK, Skala KD, Alexander BD, Huang F-L S (1990) Pituitary site of action of endothelin: selective inhibition of prolactin release in vitro. Biochem Biophys Res Comm 169:737–743
28. Stojilkovic SS, Merelli F, Iida T, Krsmanovic LZ, Catt KJ (1990) Endothelin stimulation of cytosolic calcium and gonadotropin secretion in anterior pituitary cells. Science 248: 1663–1666
29. Hirai M, Miyabo S, Ooya E, Miyanaga K, Aoyagi N, Kimura K, Kishida S, Nakai T (1991) Endothelin-3 stimulates the hypothalamic-pituitary-adrenal axis. Life Sci 48:2359–2263
30. Moretto M, Lopez FJ, Negro-Vilar A (1993) Endothelin-3 stimulates luteinizing hormone-releasing hormone (LHRH) secretion from LHRH neurons by a prostaglandin-dependent mechanism. Endocrinology 132:789–794
31. Hirata Y, Matsunaga T, Ando K, Furukawa T, Tsukagoshi H, Marumo F (1990) Presence of endothelin-1-like immunoreactivity in human cerebrospinal fluid. Biochem Biophys Res Commun 166:1274–1278
32. Yamaji T, Johshita H, Ishibashi M, Takaku F, Ohno H, Suzuki N, Matsumoto H, Fujino M (1990) Endothelin family in human plasma and cerebrospinal fluid. J Clin Endocrinol Metab 71:1611–1615
33. Kanyicska B, Burris TP, Freeman ME (1991) Endothelin-3 inhibits prolactin and stimulates LH, FSH and TSH secretion from pituitary cell culture. Biochem Biophys Res Commun 174:338–343
34. Vierhapper H, Hollenstein U, Gessl A, Roden M, Nowotny P, Waldhäusl W (1994) Effect of endothelin-1 on basal and stimulated concentrations of pituitary hormones in healthy men with and without pretreatment with nifedipine. 38th symposium, Deutsche Gesellschaft für Endokrinologie, Würzburg, 2–5 March 1994. Exp Clin Endocrinol 102 [Suppl 1]:116 (abstract)
35. Gross PM, Wainman DS, Espinosa FJ (1991) Differentiated metabolic stimulation of rat pituitary lobes by peripheral and central endothelin-1. Endocrinology 129:1110–1112
36. Calogero AE, Raiti F, Nicolosi G, Burrello N, D'Agata R, Mantero F (1994) Effects of endothelin-1 and endothelin-3 on rat hypothalamic corticotrophin-releasing hormone and pituitary ACTH release in vitro. J Endocrinol 140:419–424

Endothelin in Hypertension and Vascular Hypertrophy: Experimental Studies

E. L. Schiffrin

Introduction

Endothelins are structurally related peptides with powerful vasoconstrictor properties [1]. Endothelin-1 was initially isolated from the endothelium [1], whereas the structure of endothelin-2 and -3 was deduced from their cDNAs [2]. Endothelins are known today to be synthetized by many tissues [3–5]. In blood vessels, endothelin-1 is produced both in the endothelium [1, 2] and by smooth muscle cells [6]. Release of endothelin-1 is stimulated by low shear stress [7], hypoxia, thrombin, angiotensin II, vasopressin, norepinephrine, bradykinin and transforming growth factor β-1 (TGFβ-1) [1, 8, 9]. High shear stress is a potent inhibitor of endothelin-1 secretion [10].

Endothelin Receptors and Effects in Blood Vessels

After their release, endothelins interact with ET_A and ET_B receptors [11]. ET_A receptors are present on smooth muscle cells [12], whereas ET_B receptors [13] are present on the endothelium. Stimulation of ET_A receptors results in vascular constriction. ET_B receptor activation produces release of nitric oxide and prostacyclin, and vascular relaxation. ET_B receptors may also be involved in vasoconstriction [14], as demonstrated in aorta [15, 16], pulmonary [17] and coronary arteries [18] and in veins [19, 20] of different species. In humans, ET_B receptor-mediated constriction has been shown to occur in mammary arteries [21] and in small subcutaneous arteries [22]. ET_B receptors, however, appear to predominate in veins and pulmonary arteries, and in small arteries they seem to play a minimal role [22]. ET_B receptors in resistance arteries, however, undergo rapid desensitization on exposure to either endothelin-1 or ET_B selective agonists [23], which may result in an underestimation of their true contribution.

Endothelin-1 is not only a vasoconstrictor but also possesses mitogenic properties [24, 25]. This is another important mechanism whereby endothelins could potentially have an impact on blood pressure, by induction of vascular hypertrophy.

Plasma Levels of Endothelin-1 in Hypertension

Endothelin immunoreactivity is only slightly increased, or in many studies normal, in experimental and human hypertension [26–30]. Plasma endothelin may be increased in spontaneously hypertensive rats (SHR) treated with deoxycorticosterone acetate (DOCA) and salt, or two-kidney one clip hypertensive rats treated with caffeine [31]. In humans, hemangioendothelioma overproducing endothelin has been reported [32]. This is the only clear-cut example of increased blood-borne endothelin inducing elevated blood pressure in humans.

Vascular Reactivity to Endothelin in Hypertension

The response of the aorta of SHR to endothelin-1 is reduced [33, 34] or normal [35], and similar results have been reported in resistance arteries [34, 36, 37]. In DOCA-salt hypertensive rats responses to endothelin-1 are significantly attenuated and the density of endothelin receptors in blood vessels is reduced [38, 39]. Resistance arteries dissected from gluteal subcutaneous biopsies from hypertensive humans also present blunted media stress responses to endothelin-1 [40, 41]. Attenuated responses to endothelin, as to other agonists, are amplified as a result of the reduced vascular circumference characteristic of remodeling of small arteries in hypertension.

Vascular Production of Endothelin in Hypertension

Concentrations of endothelin-1 in plasma may not reflect vascular production of endothelin-1. When the density of endothelin receptors was found to be reduced in DOCA-salt hypertensive rats [38], we speculated that increased vascular production of endothelin in some vascular beds could account for receptor downregulation. Immunoreactive endothelin was shown to be increased by radioimmunoassay and immunohistochemistry in the endothelial layer of the vascular wall in DOCA-salt hypertensive rats (Fig. 1) [42]. Endothelin-1 mRNA was also increased in blood vessels of these hypertensive rats [43], and the only source of this increase was shown to be the endothelium by in situ hybridization histochemistry (Fig. 2) [44]. In agreement with the normal or only slightly reduced responses to endothelin and endothelin receptors in SHR blood vessels, a similar or lower content of immunoreactive endothelin was found in SHR in comparison to vessels from normotensive Wistar-Kyoto (WKY) control rats [42].

The presence of enhanced expression of endothelin-1 in blood vessels of DOCA-salt hypertensive rats, in which vascular hypertrophy is very severe [45], whereas in SHR, which do not overexpress endothelin-1 in blood vessels, vascular hypertrophy is less severe [34, 36], suggested that enhanced expression of endothelin-1 could be the cause for the difference in degree of vascular hypertrophy. If SHR are treated with DOCA and salt to induce malignant hypertension (DOCA-salt SHR), the endothelin-1 gene is overexpressed in blood vessels of SHR, and vascular hypertrophy is accentuated significantly [46].

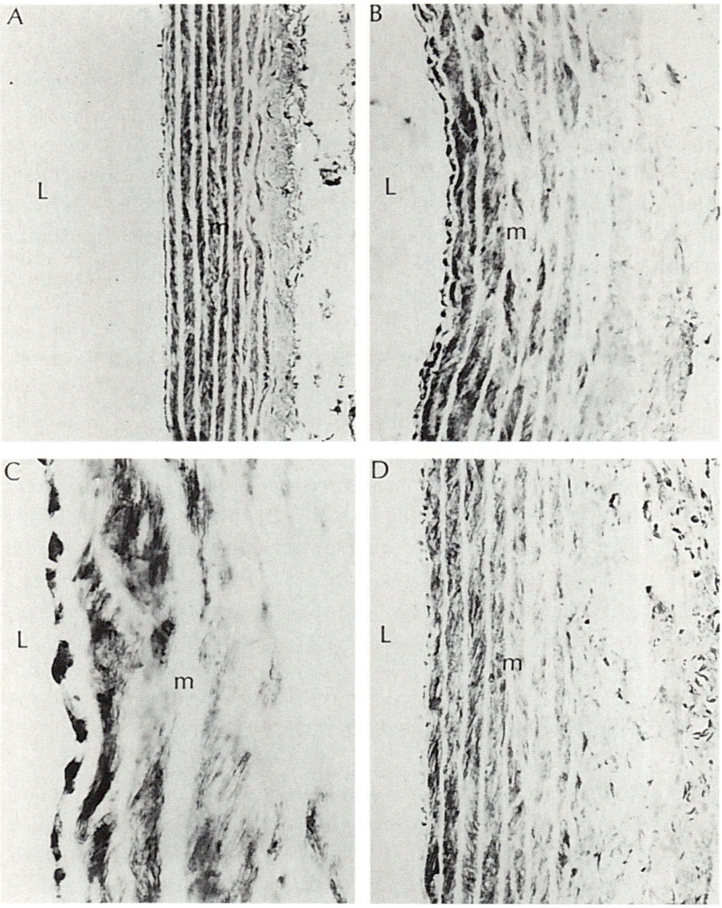

Fig. 1a–d. Photomicrographs showing immunohistochemical localization of immunoreactive endothelin-1 in aorta from uninephrectomized (**A**) and deoxycorticosterone acetate-salt hypertensive rats (**B**). A significant increase in immunostaining was detected in the endothelial cell layer of the mesenteric arteries of hypertensive rats (**B, C**). Immunostaining in endothelial cells was almost completely abolished by addition of excess synthetic endothelin-1 (**D**). Magnification of **A, B** and **D**, ×250; and **C**, ×1000. *L*, lumen; *m*, media. (From [42] with permission of the publisher)

Endothelin Antagonists and Experimental Hypertension

Administration of the selective ET_A endothelin receptor antagonist BQ-123 to rats slightly lowered blood pressure in SHR and in DOCA-salt hypertensive rats [47–49] but not in renovascular hypertension [49]. The recent availability of orally active antagonists of both ET_A and ET_B receptors [50], has allowed new insights into the implication of endothelins in blood pressure elevation and vascular hypertrophy. Acute intravenous infusion of a combined ET_A/ET_B endothelin receptor antagonist reduced blood pressure in SHR and in DOCA-salt rats [51]. Chronic oral treatement

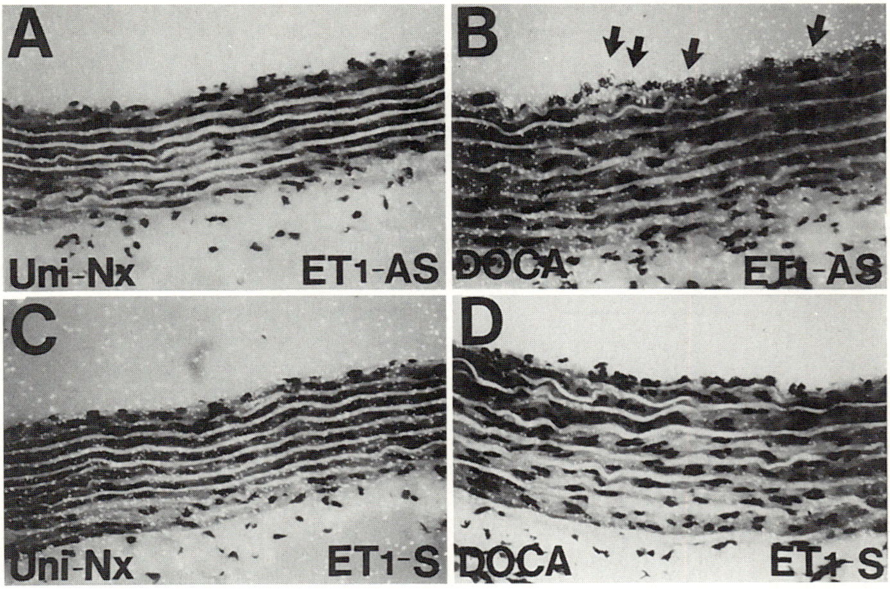

Fig. 2a–d. In situ hybridization of aorta using antisense ET-1 cRNA probes. **A** Uni-nephrectomized (Uni-Nx) control using ET-1 antisense (AS) probe, **B** Deoxycorticosterone acetate (DOCA)-salt hypertensive rat (DOCA) using ET-1 AS probe, **C** Uni-Nx control using ET-1 sense (S) probe and **D** DOCA using ET-1 S probe. Magnification, ×400. The antisense cRNA probe was radioactively labeled with both ^{35}S-UTP and ^{35}S-CTP. Note the large increase in ET-1 mRNA signal in **B** as indicated by the small arrows. The observed changes occur only in endothelial cells. (From [44] with permission of the publisher)

of DOCA-salt hypertensive rats with the combined ET_A/ET_B endothelin receptor antagonist bosentan reduced blood pressure by about 20 mmHg, and at the same time blunted the development of vascular hypertrophy and remodeling of resistance arteries (Fig. 3), beyond what could be explained by the blood pressure lowering [52]. Bosentan treatment did not affect cardiac hypertrophy in DOCA-salt hypertensive rats, which is not entirely surprising since enhanced cardiac endothelin-1 expression in DOCA-salt hypertensive rats occurs only in the endothelium of coronary arteries and the endocardium [53]. The results with chronic administration of bosentan support the hypothesis that endothelin may play a role in vascular hypertrophy in hypertension. In contrast to DOCA-salt hypertensive rats, chronic treatment with bosentan did not affect blood pressure or vascular hypertrophy in adult SHR [54] nor in young SHR [55] in agreement with the lack of vascular overexpression of endothelin-1 in these hypertensive rats. If SHR are treated with DOCA and salt, enhanced vascular expression of the endothelin-1 gene occurs and blood pressure of these malignant SHR responds now to endothelin antagonists [56, 57].

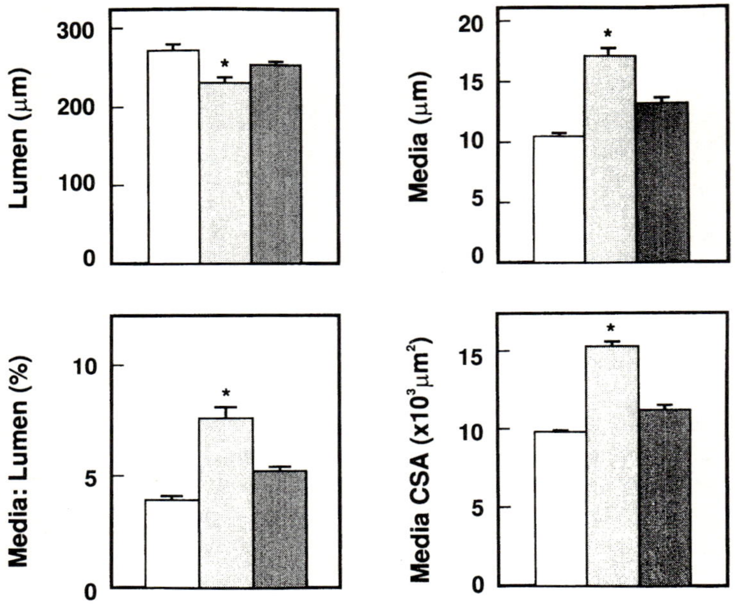

Fig. 3. Standardized parameters measured or calculated in small mesenteric arteries of uninephrectomized rats (*open column*, Uni-Nx), DOCA-salt hypertensive rats (*shaded column*, DOCA-salt), and bosentan-treated DOCA-salt hypertensive rats (solid column, DOCA-salt + Bosentan). (From [58], redrawn from data of [52] with permission of the publisher)

Conclusion

The demonstration of enhanced vascular endothelin-1 expression and the results of treatment with endothelin receptor antagonists indicate a potential involvement of endothelin-1 in the mechanisms of elevated blood pressure and severe vascular hypertrophy in DOCA-salt hypertensive rats and in DOCA-salt SHR.

Abstract

The endothelins comprise at least three 21-amino acid peptides which are potent vasoconstrictors and may also induce vascular hypertrophy. They are produced by the endothelium (particularly endothelin-1) and many other organs and tissues. The concentration of immunoreactive endothelin in plasma is normal or only slightly increased in experimental or human hypertension. Vascular responsiveness to endothelin-1 is usually normal or attenuated in hypertension. Endothelin-1 is present in increased amounts in the endothelium of blood vessels in some experimental models of hypertension such as the deoxycorticosterone acetate (DOCA)-salt hypertensive rat, in which vascular hypertrophy is much greater than in other models of hypertension. Endothelin receptor antagonist treatment attenuates partially the elevation of blood pressure and development of vascular hypertrophy in this experimental model of hypertension, whereas it has no effect on spontaneously hypertensive rats, which do not overexpress endothelin in blood vessels. These results suggest that endothelins may play a role in some experimental models of high blood pressure by accentuating hypertension-induced vascular hypertrophy.

Acknowledgements. The work of the author was supported by a Group grant from the Medical Research Council of Canada to the Multidisciplinary Research Group on Hypertension and grants from the Fondation des maladies du coeur du Québec. The author thanks Ms Angie Poliseno for secretarial help.

References

1. Yanagisawa M, Kurihara H, Kimura S, Tomobe Y, Kobayashi M, Mitsui Y, Yazaki Y, Goto K, Masaki T (1988) A novel potent vasoconstrictor peptide produced by vascular endothelial cells. Nature 332:411–415
2. Inoue A, Yanagisawa M, Kimura S, Kasuya Y, Miyauchi T (1989) The human endothelin family-3 structurally and pharmacologically distinct isopeptides predicted by 3 separate genes. Proc Natl Acad Sci USA 86:2863–2867
3. Haynes WG, Webb DJ (1993) The endothelin family of peptides: local hormones with diverse roles in health and disease? Clin Sci 84:485–500
4. Lüscher TF, Oemar BS, Boulanger CM, Hahn AWA (1993) Molecular and cellular biology of endothelin and its receptors – part I. J Hypertens 11:7–11
5. Lüscher TF, Oemar BS, Boulanger CM, Hahn AWA (1993) Molecular and cellular biology of endothelin and its receptors – part II. J Hypertens 11:121–126
6. Hahn AWA, Resink TJ, Scott-Burden T, Powell J, Dohi Y, Buhler FR (1990) Stimulation of endothelin mRNA and secretion in rat vascular smooth muscle cells: a novel autocrine function. Cell Regul 1:649–659
7. Yoshizumi M, Kurihara H, Sugiyama T, Takaku F, Yanagisawa M, Masaki T, Yasako Y (1989) Hemodynamic shear stress stimulates endothelin production by cultured endothelial cells. Biochem Biophys Res Commun 161:859–864
8. Kurihara H, Yoshizumi M, Sugiyama T, Takaku F, Yanagisawa M, Masaki T, Hamaoki M, Kato H, Yazaki Y (1989) Transforming growth factor-beta stimulates the expression of endothelin mRNA by vascular endothelial cells. Biochem Biophys Res Commun 159:1435–1440
9. Imai T, Hirata Y, Emori T, Yanagisawa M, Masaki T, Marumo F (1992) Induction of endothelin-1 gene by angiotensin and vasopressin in endothelial cells. Hypertension 19:753–757
10. Malek A, Izumo S (1992) Physiological fluid shear stress causes downregulation of endothelin-1 mRNA in bovine aortic endothelium. Am J Physiol (Cell Physiol) 263:C389–396
11. Sakurai T, Yanagisawa M, Masaki T (1 992) Molecular characterization of endothelin receptors. Trends Pharmacol Sci 13:103–108
12. Arai H, Hori S, Aramori I, Ohkubo H, Nakanishi S (1990) Cloning and expression of a cDNA encoding an endothelin receptor. Nature 348:730–732
13. Sakurai T, Yanagisawa M, Takuwa Y, Miyazaki H, Kimura S, Goto K, Masaki T (1990) Cloning of a cDNA encoding a non-isopeptide-selective subtype of the endothelin receptor. Nature 348:732–735
14. Clozel M, Gray GA, Breu V, Löffler B-M, Osterwalder R (1992) The endothelin ET_B receptor mediates both vasodilation and vasoconstriction in vivo. Biochem Biophys Res Commun 186:867–873
15. Panek RL, Major TC, Hirigorani GP, Doherty AM, Taylor DG (1992) Endothelin and structurally related analogs distinguish between endothelin receptor subtypes. Biochem Biophys Res Commun 183:566–571

16. Batra KV, McNeill JR, Xu Y, Wilson TW, Gopalkrishnan V (1993) ET_B receptors on aortic smooth muscle cells of spontaneously hypertensive rats. Am J Physiol 264:C479–C484
17. MacLean MR, McCulloch KM, Baird M (1994) Endothelin ET_A- and ET_B-receptor-mediated vasoconstriction in rat pulmonary arteries and arterioles. J Cardiovasc Pharmacol 23:838–845
18. Teerlink JR, Breu V, Sprecher U, Clozel M, Clozel J-P (1994) Potent vasoconstriction mediated by endothelin ET_B receptors in canine coronary arteries. Circ Res 74:105–114
19. Moreland SD, McMullen DM, Delaney CL, Lee VG, Hunt JT (1992) Venous smooth muscle contains vasoconstrictor ET_B-like receptors. Biochem Biophys Res Commun 184:100–106
20. Gray GA, Löffler B-M, Clozel M (1994) Characterization of endothelin receptors mediating contraction of rabbit saphenous vein. Am J Physiol (Heart Circ Physiol) 266:H959–H966
21. Seo B, Oemar BS, Siebenmann R, Von Segesser L, Lüscher TF (1994) Both ET_A and ET_B receptors mediate contraction to endothelin-1 in human blood vessels. Circulation 189:1203–1208
22. Deng LY, Li J-S, Schiffrin EL (1995) Endothelin receptor subtypes in resistance arteries from humans and from rats. Cardiovasc Res (in press)
23. Tschudi MR, Lüscher TF (1 994) Characterization of contractile endothelin and angiotensin receptors in human resistance arteries. evidence for two endothelin and one angiotensin receptor. Biochem Biophys Res Commun 204:685–690
24. Hirata Y, Takagi Y, Fukuda Y, Marumo F (1989) Endothelin is a potent mitogen for rat vascular smooth muscle cells. Atherosclerosis 78:225–228
25. Bobik A, Grooms A, Millar JA, Mitchell A, Grinpukel S (1990) Growth factor activity of endothelin on vascular smooth muscle. Am J Physiol (Cell Physiol) 258:C408–C415
26. Suzuki N, Miyauchi T, Tomobe Y, Matsumoto H, Goto K, Masaki T, Fujino M (1990) Plasma concentrations of endothelin-1 in spontaneously hypertensive rats and DOCA-salt hypertensive rats. Biochem Biophys Res Commun 167:941–947
27. Kohno M, Yasumari K, Murakawa Kl, Yokokawa K, Horio T, Fukui T, Takeda T (1990) Plasma immunoreactive endothelin in essential hypertension. Am J Med 88:614–618
28. Davenport AP, Ashby MJ, Easton P, Ella S, Bedford J, Dickerson C, Nunez DJ, Capper SJ, Brown MJ (1990) A sensitive radioimmunoassay measuring endothelin-like immunoreactivity in human plasma: comparison of levels in patients with essential hypertension and normotensive control subjects. Clin Sci 78:261–264
29. Shichiri M, Hirata Y, Ando K, Emori T, Ohta K, Kimoto S, Ogura M, Inoue, A, Marumo F (1990) Plasma endothelin levels in hypertension and chronic renal failure. Hypertension 15:493–496
30. Schiffrin EL, Thibault G (1991) Plasma endothelin in human essential hypertension. Am J Hypertens 4:303–308
31. Kohno M, Murakawa K-i, Horio T, Yokokawa K, Yasunari K, Fukui T, Takeda T (1991) Plasma immunoreactive endothelin-1 in experimental malignant hypertension. Hypertension 18:93–100
32. Yokokawa K, Tahara H, Kohno M, Murakawa Ki, Yasunari K, Nakagawa K, Hamada T, Otani S, Yanagisawa M, Takeda T (1991) Hypertension associated with endothelin-secreting malignant hemangioendothelioma. Ann Intern Med 114:213–215
33. Clozel M (1989) Endothelin sensitivity and receptor binding in the aorta of spontaneously hypertensive rats. J Hypertens 17:913–917
34. Nguyen PV, Yang X-P, Li G, Deng LY, Flückiger J-P, Schiffrin EL (1993) Contractile responses and signal transduction of endothelin-1 in aorta and mesenteric vasculature of adult spontaneously hypertensive rats. Can J Physiol Pharmacol 71:473–483
35. Bolger GT, Liard F, Jodoin A, Jaramillo J (1991) Vascular reactivity, tissue levels and binding sites for endothelin: a comparison in the spontaneously hypertensive and Wistar-Kyoto rats. Can J Physiol Pharmacol 69:406–413
36. Deng LY, Schiffrin EL (1992) Effects of endothelin-1 and vasopressin on resistance arteries of spontaneously hypertensive rats. Am J Hypertens 5:817–822
37. Dohi Y, Lüscher TF (1991) Endothelin in hypertensive resistance arteries: intraluminal and extraluminal dysfunction. Hypertension 18:543–549
38. Nguyen PV, Parent A, Deng LY, Flückiger JP, Thibault G, Schiffrin EL (1992) Endothelin vascular receptors and responses in DOCA-salt hypertensive rats. Hypertension 19 [Suppl I]:II-98–II-104
39. Flückiger JP, Nguyen PV, Li G, Yang XP, Schiffrin EL (1992) Phosphoinositide, calcium and 1,2 diacylglycerol response of blood vessels of DOCA-salt hypertensive rats to endothelin-1. Hypertension 19:743–749

40. Schiffrin EL, Deng LY, Larochelle P (1992) Blunted effects of endothelin on small subcutaneous resistance arteries of mild essential hypertensive patients. J Hypertens 10:437–444
41. Schiffrin EL, Deng LY, Larochelle P (1993) Morphology of resistance arteries and comparison of effects of vasoconstrictors in mild essential hypertensive patients. Clin Invest Med 16:177–186
42. Larivière R, Thibault G, Schiffrin EL (1993) Increased endothelin-1 content in blood vessels of deoxycorticosterone acetate-salt hypertensive but not in spontaneously hypertensive rats. Hypertension 21:294–300
43. Larivière R, Day R, Schiffrin EL (1993) Increased expression of endothelin-1 gene in blood vessels of deoxycorticosterone acetate-salt hypertensive rats. Hypertension 21:916–920
44. Day R, Larivière R, Schiffrin EL (1995) In situ hybridization shows increased endothelin-1 mRNA levels in endothelial cells of blood vessels of deoxycorticosterone acetate-salt hypertensive rats. Am J Hypertens 8 (in press)
45. Deng LY, Schiffrin EL (1992) Effects of endothelin on resistance arteries of DOCA-salt hypertensive rats. Am J Physiol 262:H1782–H1787
46. Schiffrin EL, Larivière R, Li J-S, Sventek P, Touyz RM (1995) Deoxycorticosterone acetate plus salt induce overexpression of vascular endothelin-1 and severe vascular hypertrophy in spontaneously hypertensive rats. Hypertens 25 [Part 2]:769–773
47. McMahon EG, Palomo MA, Brown MA, Bertenshaw ST, Carter JS (1993) Effect of phosphoramidon (endothelin converting enzyme inhibitor) and BQ-123 (endothelin receptor subtype A antagonist) on blood pressure in hypertensive rats. Am J Hypertens 6:667–673
48. Bazil MK, Lappe RW, Webb RL (1992) Pharmacologic characterization of an endothelina (ET_A) receptor antagonist in conscious rats. J Cardiovasc Pharmacol 20:940–948
49. Nishikibe M, Tsuchida S, Okada M, Fukuroda T, Shimamoto K, Yano M, Ishikawa K, Ikemoto F (1993) Antihypertensive effect of newly synthesized endothelin antagonist, BQ-123, in a genetic hypertensive model. Life Sci 52:717–724
50. Clozel M, Breu V, Burri K, Cassal J-M, Fischli W, Gray GA, Hirth G, Löffler, Müller M, Neidhart W, Ramuz H (1993) Pathophysiological role of endothelin revealed by the first orally active endothelin receptor antagonist. Nature 365:759–761
51. Clozel M, Clozel JP, Hess P (1993) Endothelin receptor antagonism: a new therapeutic approach in experimental hypertension. Circulation 88 [Suppl]:I-316 (abstract)
52. Li J-S, Larivière R, Schiffrin EL (1994) Effect of a nonselective endothelin antagonist on vascular remodeling in DOCA-salt hypertensive rats. Evidence for a role of endothelin in vascular hypertrophy. Hypertension 24:183–188
53. Larivière R, Deng LY, Day R, Sventek P, Thibault G, Schiffrin EL (1995) Increased endothelin-1 expression in the endothelium of coronary arteries and endocardium in the DOCA-salt hypertensive rat. J Mol Cell Cardiol in press
54. Li J-S, Schiffrin EL (1995) Effect of chronic treatment with a combined ET_A/ET_B endothelin receptor antagonist in adult spontaneously hypertensive rats. Hypertension 25 [Part 1]:495–500
55. Li J-S, Schiffrin EL (1995) Chronic endothelin receptor antagonist treatment of young spontaneously hypertensive rats. J Hypertension 13:in press
56. Okada M, Fukuroda T, Shimamoto K, Takahashi R, Ikemoto F, Yano M, Nishikibe M (1994) Antihypertensive effects of BQ-123, a selective endothelin ET_A receptor antagonist, in spontaneously hypertensive rats treated with DOCA-salt. Eur J Pharmacol 259:339–342
57. Schiffrin EL, Sventek P, Li J-S, Turgeon A, Reudelhuber T (1995) Antihypertensive effect of bosentan, a mixed ET_A/ET_B endothelin receptor antagonist, in DOCA-salt spontaneously hypertensive rats. Brit J Pharmacol in press
58. Schiffrin EL (1995) Endothelin, potential role in hypertension and vascular hypertrophy. Brief invited review. Hypertension 25:1135–1143

Role of Endogenous Endothelin in Myocardial Ischemia and Nitric Oxide-Deficient Hypertension

V. Richard, M. Hogie, and C. Thuillez

Introduction

Vascular endothelial cells synthesize various vasoactive substances, including endothelin-1 (ET-1), a 21 amino acid vasoconstrictor peptide [54]. Although experiments have suggested that ET-1 is released in basal conditions [1, 45], the exact role of this vasoconstrictor peptide in the control of vascular tone in vivo remains elusive. Experimental and clinical evidence suggests that the release as well as the biological actions of ET-1 are increased in various pathological situations, such as myocardial infarction or heart failure. Furthermore, there is also evidence that both the synthesis and the vasoconstrictor effects of ET-1 are negatively regulated through continuous release of inhibitory factors, such as nitric oxide (NO). Thus, it is possible that the biological effects of endogenous ET-1 might be evidenced only in pathological conditions, or in situations of impaired NO synthesis. This article reviews current knowledge on the potential role of endogenous ET-1 in myocardial ischemia and infarction, as well as the role of this peptide in acute situations of impaired NO synthesis, especially focusing on recent evidence obtained in our laboratory with the use of specific antagonists of ET-1 receptors.

Endothelin in Myocardial Ischemia and Infarction

Myocardial and Coronary Effects of Exogenous Endothelin

Shortly after the discovery of ET-1, observation of the biological effects of exogenously administered ET-1 made this peptide a potential contributor to ischemia and reperfusion-induced injury. Indeed, in vivo administration of ET-1 markedly increases arterial blood pressure [54]. Furthermore, local administration of ET-1 induces a marked coronary vasoconstriction in various species, including dogs [8, 12, 23], rabbits [16, 28], rats [2, 14, 32], pigs [13], and man [6, 7, 46, 55]. In some experiments, these coronary vasoconstrictor effects are accompanied by reductions in cardiac contractility, release of myocardial lactate and ST segment elevation, which are signs of the development of myocardial ischemia [13, 16, 17, 23].

In addition to being a potent vasoconstrictor, ET-1 exerts positive chronotropic and inotropic effects [18, 19, 25, 42, 43]. ET-1 may also be arrhythmogenic [39, 57].

Interestingly, several experiments suggest that the biological responses to exogenous ET-1 (and especially the coronary constrictor effects) are increased during myocardial ischemia and reperfusion [5, 14, 33, 34, 53]. Furthermore, ischemia may result in an externalization of ET-1 receptors in cardiac membranes, indicating that the direct actions of this peptide on the myocyte may be increased in ischemic conditions [26]. Indeed, ET-1 pretreatment reduces the time to contracture in isolated, globally ischemic rat hearts, suggesting a proischemic effect independent of changes in coronary blood flow [14].

Release of ET-1 During Myocardial Ischemia and Reperfusion

Numerous experimental studies demonstrated that myocardial ischemia and reperfusion increase plasma levels of ET-1 [12, 47, 48, 52]. An increased level of plasma ET-1 has also been shown after acute myocardial infarction in humans [31, 44, 56]. However, in most of these studies, it is difficult to assess whether such increases are indeed a direct consequence of myocardial ischemia and reperfusion, or are at least in part an indirect consequence of the ischemia-induced hemodynamic changes. However, recent experiments performed in dogs show that ischemia and reperfusion stimulate the release of ET-1 from the ischemic myocyte [49], suggesting that myocardial ischemia may per se be a stimulus for ET-1 synthesis.

Endothelin Antagonists and Myocardial Ischemia/Reperfusion Injury

Given the pharmacological profile of ET-1, as well as the observation that ischemia and reperfusion are associated with an increased production of ET-1, it is tempting to conclude that endogenous ET-1 may play a pathophysiological role in situations of myocardial ischemia and reperfusion, either through coronary constriction (possibly by decreasing the recovery of blood flow during reperfusion and thus aggravating the "no-reflow" phenomenon), or through direct myocardial effects (such as increased oxygen demand of increased intracellular calcium). Indeed, studies performed in rats or rabbits have suggested that monoclonal antibodies directed against ET-1, or phosphoramidon (an inhibitor of endothelin converting enzyme) could reduce ischemic damage in experimental models of myocardial ischemia and reperfusion [14, 22, 52]. A beneficial effect has also been reported in dogs with the specific ET_A antagonist BQ-123 [15]. In contrast, in a rabbit model, the ET_A antagonist FR 139317, administered at doses which markedly the pressor effects of exogenous ET-1 [28], did not affect ischemia-induced ST segment elevation, nor did it affect infarct size [29]. Moreover, preliminary data obtained in dogs show that intracoronary administration of FR 139317 or the non selective endothelin receptor antagonist PD 145065 both induce a significant increase in infarct size [50, 51], suggesting that endogenous ET-1 might be protective in this model. Thus, the role of endogenous ET-1 in myocardial ischemia/reperfusion injury remains controversial (Table 1).

To further assess this role, we performed experiments using bosentan [10], a non peptide antagonist of the two known subtypes of ET receptors (ET_A and ET_B), in pentobarbital-anesthetized rats subjected to left coronary artery occlusion followed by reperfusion. Specifically, the effect of bosentan was investigated on three major

Table 1. Studies on the effect of anti-endothelin antibodies, endothelin converting enzyme inhibitors (phosphoramidon), or endothelin receptor antagonists on infarct size in animal models of myocardial ischemia/reperfusion

Drug	Species	Infarct size	Study
Antibody	Rat	Decreased	Watanabe et al. [52]
Antibody	Rabbit	Decreased	Kusumoto et al. [22]
Phosphoramidon	Rat	Decreased	Grover et al. [14]
ETA antagonist (BQ-123)	Dog	Decreased	Grover et al. [15]
ETA antagonist (FR 139 317)	Rabbit	No effect	McMurdo et al. [29]
ET_A-ET_B antagonist (Bosentan)	Rat	No effect	Richard et al. [36]
ETA antagonist (FR 139 317)	Dog	Increased	Velasco et al. [50]
ET_A-ET_B antagonist (PD 145 065)	Dog	Increased	Velasco et al. [51]

aspects of ischemia/reperfusion injury, i.e. reperfusion-induced arrhythmias, infarct size, and reperfusion-induced coronary endothelial dysfunction [36].

In this model, reperfusion after a 5-min period of ischemia was associated with the occurrence of ventricular fibrillation in 86% of the animals; fibrillation was irreversible in 21% of the animals. Bosentan did not affect the incidence of total of irreversible reperfusion-induced ventricular fibrillation (Fig. 1). In the same model a more prolonged coronary artery occlusion (20 min), followed by reperfusion (1 h) resulted in the development of myocardial infarcts (assessed using triphenyl tetrazolium staining) involving 63% ± 4% of the area at risk. In those experimental conditions, bosentan also did not affect infarct size (60% ± 6% or the area at risk; Fig. 2).

The effect of ischemia and reperfusion and of bosentan on coronary endothelial responses were studied on coronary artery segments (internal diameter 250–300 μm)

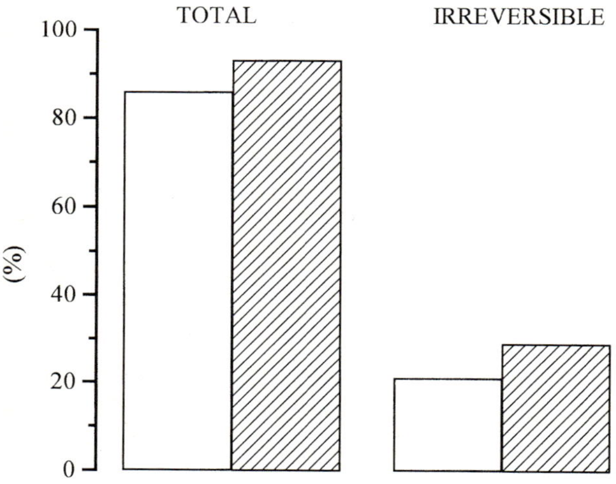

Fig. 1. Incidence of total and irreversible ventricular fibrillation upon reperfusion after a 5-min period of ischaemia in control rats (*open bars*; $n=14$) or in rats treated by bosentan (3 mg/kg; *hatched bars*; $n=15$). Bosentan did not affect the incidence of severe reperfusion-induced arrhythmias. (From [36])

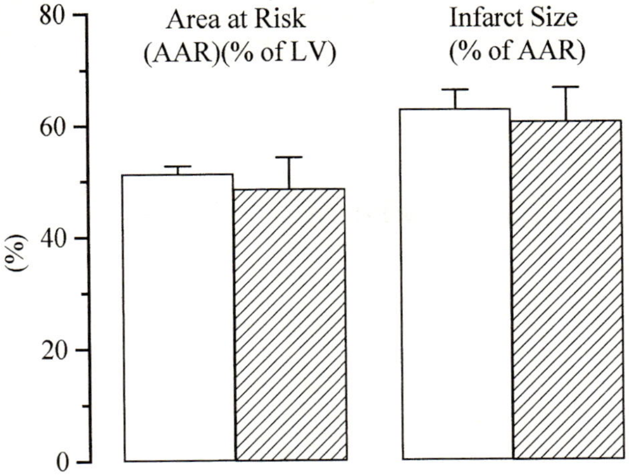

Fig. 2. Area at risk (AAR), expressed as percentage of left venricle (LV), and infarct size, expressed as percentage of AAR, in controls rats (*open bars*; $n=10$), or in animals treated by bosentan (3 mg/kg; *hatched bars*; $n=8$). Bosentan did not affect infarct size. (From [36])

taken distal to the occlusion site and mounted in small vessel myographs. In arteries taken from sham-operated rats (precontracted with serotonin), acetylcholine induced endothelium-dependent relaxations, which were markedly reduced after ischemia and reperfusion, in agreement with our previous results [37]. This reperfusion-induced coronary endothelial dysfunction was not affected by bosentan (Fig. 3).

Thus, these experiments performed in rats suggest that the administration of a mixed antagonist of ET_A and ET_B receptors bosentan did not have any significant

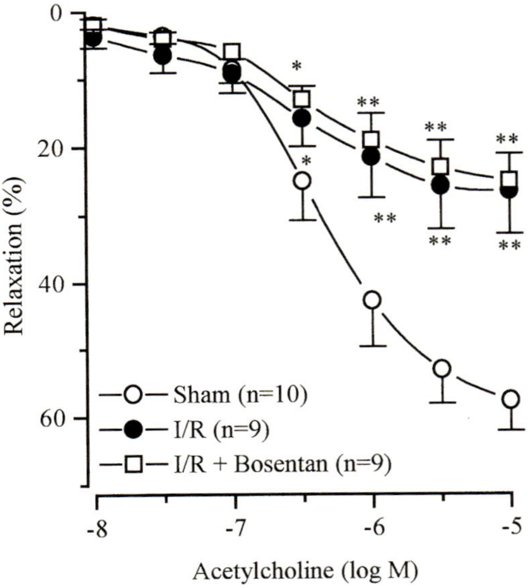

Fig. 3. Relaxing responses of rat corona arteries to acetylcholine. Arteries were taken from sham-operated animals (*open circle*; $n=10$), or animals subjected to ischaemia followed by reperfusion (I/R) in the absence (*solid circle*; $n=9$) or in the presence of bosentan (*open square*; $n=9$). Values are expressed as a percentage of the contraction to serotonin. * $p<0.05$; ** $p<0.01$ vs sham. The endothelium-dependent relaxations to acetylcholine were markedly reduced after ischemia and reperfusion, and this was not affected by bosentan, suggesting that this drug did not affect reperfusion-induced coronary endothelial dysfunction. (From [36])

effect on infarct size, reperfusion-induced arrhythmias or reperfusion-induced coronary endothelial dysfunction. These results do not support the hypothesis that endogenous ET-1 contributes to myocyte or coronary endothelial injury after ischemia and reperfusion.

Endothelin in Nitric Oxide-Deficient Hypertension

As mentioned above, the exact role of endogenous ET-1 in the control of vascular tone in vivo remains elusive. Indeed, specific antagonists of endothelin receptors do not induce hypotension in normal animals [9, 35]. This may be due in part to the fact that the continuous production of NO from endothelial cells permanently inhibits the synthesis and/or the vasoconstrictor effects of ET-1 (Fig. 4). Indeed, NO donors or agents that increase the production of NO from endothelial cells decrease the cellular production of ET-1, whereas inhibition of NO synthesis increases the release of the peptide [3, 4, 21, 40]. In addition, exogenous or endogenous NO fully reverse contractions induced by ET-1 [11, 20, 27, 30, 54], whereas endothelium removal or inhibition of basal release of NO potentiates the vasoconstrictor responses to this peptide [24, 27, 41].

Thus, the vasoconstrictor effects of ET-1 might be evidenced only in situations of impaired NO synthesis. If this were true, then it is possible that part of the hypertensive effect induced by inhibitors of NO synthesis would be in fact the consequence of an unmasking of a vasomotor tone due to ET-1 (Fig. 4).

To address this question, we designed experiments to test whether the pressor effects induced in anesthetized rats by the L-arginine analogue N^G-nitro L-arginine methyl ester (L-NAME) were affected by the mixed ET_A/ET_B receptor antagonist

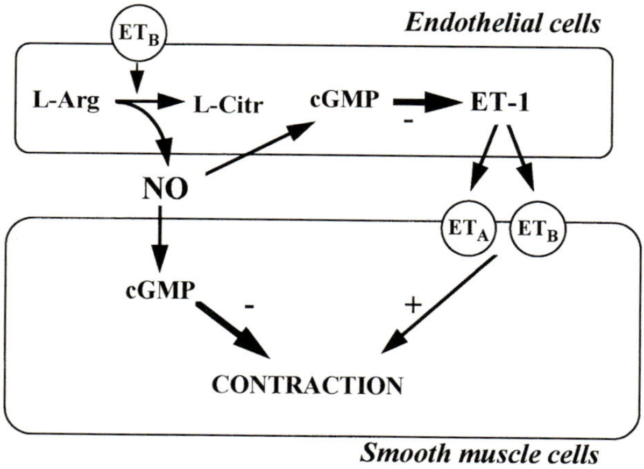

Fig. 4. Possible interactions between endothelin-1 (*ET-1*) and nitric oxide (*NO*) in the vascular wall. Nitric oxide is a potent inhibitor of endothelin-induced contractions and may also inhibit the endothelial release of the peptide. *L-Arg*, L-arginine; *L-Citr*, L-citrulline; cGMP, 3′-5′ cyclic guanosine monophosphate

bosentan [38]. For this purpose, rats were instrumented with a carotid Millar micromanometer for the measurement of arterial pressure, and a Doppler probe placed around the ascending aorta in order to measure cardiac output.

In this preparation (Fig. 5), L-NAME induced a marked increase in mean arterial pressure. This pressor response appeared entirely mediated through an increase in total vascular resistance, since cardiac output decreased significantly after L-NAME. L-NAME also induced a significant decrease in heart rate, stroke volume or cardiac index.

In basal conditions, (i.e. in the absence of L-NAME), bosentan did not affect arterial pressure, heart rate or cardiac output. However, bosentan markedly reduced the pressor response to L-NAME. This effect appeared entirely the consequence of a lesser increase in total peripheral resistance, since bosentan significantly blunted the L-NAME-induced decreases in cardiac output, stroke volume or cardiac index, without affecting its effect on heart rate.

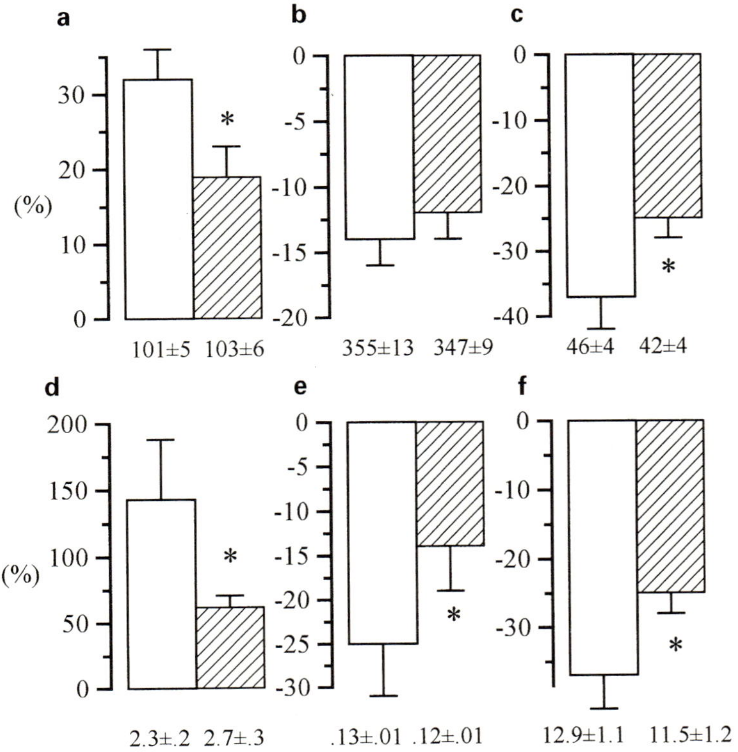

Fig. 5 a–f. Mean ± SEM percent changes in: **a** mean arterial pressure, **b** heart rate, **c** cardiac output, **d** total peripheral resistance, **e** stroke volume, and **f** cardiac index induced by L-NAME (3 mg/kg) in the absence (control, *open bars*; $n=12$) or in the presence of bosentan (3 mg/kg; *hatched bars*; $n=12$). *Values* under the individual graphs are the absolute values of the corresponding parameters measured immediately before administration of L-NAME (i.e. 15 min after administration of saline or bosentan). * $p<0.05$ vs control

Thus, our experiments demonstrate that part of the hypertensive response induced by L-NAME is indeed due to unmasking of an ET-1-induced vasopressor tone (Fig. 4), which, based on our previous findings [38], appears mostly mediated through stimulation of the smooth muscle ET_A receptors, and could be at least in part the consequence of an increased production of ET-1 [38]. This experiments suggest that ET-1 may play a major role in pathophysiological situations associated with an impaired release of NO.

Abstract

Although it is clear that endothelial cells produce endothelin-1 (ET-1), the role of this vasoconstrictor peptide in the regulation of cardiovascular function remains elusive. There is evidence, however, that the release and biological effects of ET-1 may be increased in various pathological situations, such as myocardial ischemia and reperfusion, as well as after inhibition of endothelial synthesis of nitric oxide (NO). This article reviews current knowledge on the potential role of ET-1 in myocardial ischemia and infarction as well as the role of this peptide in NO-deficient hypertension.

During myocardial ischemia, initial evidence of the potent coronary constrictor effects of ET-1 made this peptide a potential contributor to myocardial ischemia and reperfusion injury. Furthermore, experimental and clinical studies showed that plasma levels of ET-1 were elevated during myocardial ischemia and infarction. Recently, the potential role of ET-1 in myocardial ischemia-reperfusion injury has been assessed using newly developed ET receptor antagonists. However, results in myocardial ischemia have been controversial. Indeed, although initial studies with anti-ET antibodies and endothelin converting enzyme inhibitors reported positive effects during ischemia, more recent studies with specific antagonists of ET receptors failed to detect beneficial effects in animal models of ischemia. For example, we showed in rats that the nonpeptide ET antagonist bosentan failed to affect various aspects of ischemia/reperfusion injury, such as reperfusion arrhythmias, infarct size, or reperfusion-induced coronary endothelial dysfunction. Thus, these results do not support the hypothesis that ET-1 contributes to myocyte or coronary endothelial injury after ischemia and reperfusion.

In contrast, our results obtained in a rat model of NO-deficient hypertension indeed show that the pressor responses induced by inhibitors of NO synthesis were markedly reduced by ET antagonists such as bosentan or BQ-123, suggesting that part of NO-deficient hypertension was due to an unmasking of ET-induced vasopressor responses. Thus, these results, together with previous in vitro and in vivo data which showed that basal release of NO continuously inhibits both the release and the vasoconstrictor effects of ET, suggest that this peptide may play a major role in pathophysiological situations associated with an impaired release of NO.

References

1. Ando K, Hirata Y, Schichiri M, Emori T, Maruno F (1989) Presence of immunoreactive endothelin in human plasma. FEBS Lett 245:164–166
2. Baydoun BR, Peers SH, Cirino G, Woodward B (1989) Effects of endothelin-1 on the rat isolated heart. J Cardiovasc Pharmacol 13 [Suppl 5]:S193–S196
3. Boulanger C, Lüscher TF (1990) Release of endothelin from the porcine aorta: inhibition by endothelium-derived nitric oxide. J Clin Invest 85:587–590
4. Boulanger CM, Lüscher TF (1991) Hirudin and nitrates inhibit the thrombin-induced release of endothelin from the intact porcine aorta. Circ Res 68:1768-1772
5. Brunner F, du Toit EF, Opie LH (1992) Endothelin release during ischaemia and reperfusion of isolated perfused rat hearts. J Mol Cell Cardiol 24:1291–1305
6. Chester AH, Dashwood MR, Clarke JG, Larkin SW, Davies GJ, Tadjkarimi S, Maseri A, Yacoub MH (1989) Influence of endothelin on human coronary arteries and localization of its binding sites. Am J Cardiol 63:1395–1398
7. Clarke JG, Benjamin N, Larkin SW, Webb DJ, Keogh BE, Davies GJ, Maseli A (1989) Endothelin is a potent long-lasting vasoconstrictor in man. Am J Physiol 257:H2033–H2035
8. Clozel JP, Clozel M (1989) Effects of endothelin on the coronary vascular bed in open-chest dogs. Circ Res 65:1193–1200
9. Clozel M, Breu V, Burri K, Cassal JM, Fischli W, Gray GA,Hirth G, Löffler BM, Müller M, Neidhart W, Ramuz H (1993) Pathophysiological role of endothelin revealed by the first orally active endothelin receptor antagonist. Nature 365:759–761
10. Clozel M, Breu V, Gray GA, Kalina B, Löffler B, Burri K, Cassal J, Hirth G, Müller M, Neidhart W, Ramuz H (1994) Pharmacological characterization of bosentan, a new potent orally active nonpeptide endothelin receptor antagonist. J Pharmacol Exp Ther 270:228–235
11. Dohi Y, Lüscher TF (1991) Endothelin in hypertensive resistance arteries. Intraluminal and extraluminal dysfunction. Hypertension 18:543–549
12. Donckier J, Hanet C, Stoleru L, van Mechelen HV, Galanti L, Hayashida W, Keyeux A, Keteslegers J-M, Pouleur H (1994) Effects of endothelin-1 at pathophysiological concentrations on coronary perfusion and mechanical function of normal and postischemic myocardium. J Cardiovasc Pharmacol 23:212–219
13. Ezra D, Goldstein RE, Czaja JF, Feuerstein GZ (1989) Lethal ischemia due to intracoronary endothelin in pigs. Am J Physiol 257:H339–H343
14. Grover GJ, Sleph PG, Fox M, Trippodo NC (1992) Role of endothelin-1 and big endothelin-1 in modulating coronary vascular tone, contractile function and severity of ischemia in rat hearts. J Pharmacol Exp Ther 263:1074–1082
15. Grover GJ, Dzwonczyk S, Parham CS (1993) The endothelin-1 receptor antagonist BQ-123 reduces infarct size in a canine model of coronary occlusion and reperfusion. Cardiovasc Res 27: 1613–1618
16. Hirata K, Matsuda Y, Akita H, Yokoyama M, Fukuzaki H (1990) Myocardial ischemia induced by endothelin in the intact rabbit: angiographic analysis. Cardiovasc Res 24:879–883
17. Hori S, Kyotani S, Inoue S, Fukuda K, Ohnishi Y, Kusuhara M, Aikawa N, Yamaguchi K, Nakamura Y, Handa S (1991) Subepicardial microischemia formation induced by epicardial application of endothelin-1. J Cardiovasc Pharmacol 17 [Suppl 7]:S300–S301
18. Hu JR, von Hardsdorf R, Lang RE (1988) Endothelin has potent inotropic effects in rat atria. Eur J Pharmacol 158:275–278
19. Ishikawa T, Yanagisawa M, Kimura S, Goto K, Masaki T (1988) Positive inotropic action of novel vasoconstrictor peptide endothelin on guinea-pig atria. Am J Physiol 255:H970–H973
20. Ito S, Juncos LA, Nushiro N, Johnson CS, Carretero OA (1991) Endothelium-derived relaxing factor modulares endothelin action in afferent arterioles. Hypertension 17:1052–1056
21. Kourembanas S, McQuillan LP, Leung GK, Faller DV (1993) Nitric oxide regulates the expression of vasoconstrictors and growth factors by vascular endothelium under both normoxia and hypoxia. J Clin Invest 92:99–104
22. Kusumoto K, Awane Y, Fujiwara S, Watanabe T (1993) Role of endogenous endothelin in extension of rabbit myocardial infarction. J Cardiovasc Pharmacol 22 [Suppl 8]:S339–S342

23. Larkin SW, Clarke JG, Keogh BE, Araujo L, Rhodes C, Davies GJ, Taylor KM, Maseri A (1989) Intracoronary endothelin induces myocardial ischemia by small vessel constriction in the dog. Am J Cardiol 64:956–958
24. Lerman A, Sandok EK, Hilderbrand FL, Burnett JC Jr (1992) Inhibition of endothelium-derived relaxing factor enhances endothelin-mediated vasoconstriction. Circulation 85:1894–1898
25. Li K, Stewart DJ, Rouleau JL (1991) Myocardial contractile actions of endothelin-1 in rat and rabbit papillary muscles. Role of endocardial endothelium. Circ Res 69:301–312
26. Liu J, Chen R, Casley DJ, Nayler WG (1990) Ischemia and reperfusion increase 125I-labeled endothelin-1 binding in rat cardiac membranes. Am J Physiol 258:H829–H835
27. Lüscher TF, Yang Z, Von Segesser L, Stulz P, Boulanger C, Siebenmann R, Turina M, Buhler F (1990) Interaction between endothelin- 1 and endothelium-derived relaxing factor in human arteries and veins. Circ Res 66:1088–1094
28. McMurdo L, Corder R, Thiemermann C, Vane JR (1993) Incomplete inhibition of the pressor effect of endothelin-1 and related peptides in the anaesthetized rat with BQ-123 provides evidence for more than one vasoconstrictor substance. Br J Pharmacol 108:557–561
29. McMurdo L, Thiemermann C, Vane JR (1994) The effects of the endothelin ETA receptor antagonist, FR 139 317, on infarct size in a rabbit model of acute myocardial ischemia and reperfusion. Br J Pharmacol 112:75–80
30. Miller VM, Komori K, Burnett JCJ, Vanhoutte PM (1989) Differential sensitivity to endothelin in canine arteries and veins. Am J Physiol 257:H1127–H1131
31. Miyauchi T, Yanagisawa M, Tomizawa T, Sugishita Y, Suzuki N, Fujino M, Ajisaka R, Goto K, Masaki T (1989) Increased plasma concentrations of endothelin-1 and big endothelin-1 in acute myocardial infarction (letter). Lancet 2:53–54
32. Neubauer S, Ertl G, Haas U, Pulzer F, Kocksiek K (1990a) Effects of endothelin-1 in isolated perfused rat heart. J Cardiovasc Pharmacol 16:1–8
33. Neubauer S, Ertl G, Pulzer F, Haas U, Ehrsch A, Zimmerman S, Kocksiek K (1990b) Effects of endothelin-1 in the isolated heart under ischemic and cardioplegic conditions. J Cardiovasc Pharmacol 16:804–811
34. Neubauer S, Zimmermann S, Hirsch A, Pulzer F, Tian R, Bauer W, Bauer B, Ertl G (1991) Effects of endothelin-1 in the isolated heart in ischemia/reperfusion and hypoxia/reoxygenation injury. J Mol Cell Cardiol 23:1397–1409
35. Qhlstein EH, Douglas SA, Ezekiel M, Gellai M (1993) Antihypertensive effects of the endothelin receptor antagonist BQ-123 in conscious spontaneously hypertensive rats. J Cardiovasc Pharmacol 22 [Suppl 8]:S321-S324
36. Richard V, Kaeffer N, Hogie M, Tron C, Blanc T, Thuillez C (1 994) Role of endogenous endothelin in myocardial and coronary endothelial injury after ischemia and reperfusion in rats: studies with bosentan, a mixed ETA-ETB antagonist. Br J Pharmacol 113:869–876
37. Richard V, Kaeffer N, Tron C, Thuillez C (1994) Ischemic preconditioning protects against coronary endothelial dysfunction induced by ischemia and reperfusion. Circulation 89:1254–1261
38. Richard V, Hogie M, Clozel M, Lîffler B, Thuillez C (1995) In vivo evidence of an endothelin-induced vasopressor tone after inhibition of nitric oxide synthesis in rats. Circulation 91: 771–775
39. Salvati P, Chierchia S, Dho L, Ferrario RG, Parenti P, Vicedomini G, Patrono C (1991) Proarrhythmic activity of intracoronary endothelin in dogs: relation to the site of administration and to changes in regional flow. J Cardiovasc Pharmacol 17:1007–1014
40. Saijonmaa O, Ristimaki A, Fyhrquist F (1990) Atrial natriuretic peptide, nitroglycerine, and nitroprusside reduce basal and stimulated endothelin production from cultured endothelial cells. Biochem Biophys Res Commun 173:514–520
41. Seo B, Oemar BS, Siebenmann R, Von Segesser L, Lüscher TF (1994) Both ETA and ETB receptors mediate contraction to endothelin-1 in human blood vessel. Circulation 89:1203–1208
42. Shah AM, Lewis MJ, Henderson AH (1989) Inotropic effects of endothelin in ferret ventricular myocardium. Eur J Pharmacol 163:365–367
43. Shomisch Moravec C, Reynolds EE, Stewart RW, Bond M (1989) Endothelin is a positive inotropic agent in human and rat heart in vitro. Biochem Biophys Res Commun 159:14–18
44. Stewart DJ, Kubac G, Costello KB, Cemacek P (1991) Increased plasma endothelin-1 in the early hours of acute myocardial infarction. J Am Coll Cardiol 18:38–43

45. Suzuki N, Matsumoto H, Kitada C (1989) Immunoreactive endothelin-1 in plasma detected by a sandwich-type enzyme immunoassay. J Cardiovasc Pharmacol 13:151–153
46. Tippins JR, Antoniw JW, Maseri A (1989) Endothelin-1 is a potent constrictor in conductive and resistive coronary arteries. J Cardiovasc Pharmacol 13 [Suppl 5]: S177–S179
47. Tsuji S, Sawamura A, Watanabe H, Takihara K, Park SE, Azuma J (1991) Plasma endothelin levels during myocardial ischemia and reperfusion. Life Sci 48:1745–1749
48. Velasco CE, Jackson EK, Morrow JA, Vitola JV, Inagami T, Forman MB (1993) Intravenous adenosine suppresses cardiac release of endothelin after myocardial ischaemia and reperfusion. Cardiovasc Res 27:121–128
49. Velasco C, Giaid A, Williamson J, Yanagisawa M (1994) Myocardial ischemia and local endothelin-1 production. Eur Heart J 15 [Suppl]:118
50. Velasco CE, Yanagisawa M, Williamson JL, Femando Triana J (1993) Cardioprotective action of endothelin-1 during ischemia. Circulation 88 (2):I-544
51. Velasco CE, Yanagisawa M (1994) Endothelins protect the ischemic myocardium. J Am Coll Cardiol [Suppl]:30A
52. Watanabe T, Suzuki N, Shimamoto N, Fujino M, Imada A (1991) Contribution of endogenous endothelin to the extension of myocardial infarct size in rats. Circ Res 69:370–377
53. Watts JA, Chapat S, Johnson DE, Janis RA (1992) Effects of nisoldipine upon vasoconstrictor responses and binding of endothelin-1 in ischemic and reperfused rat hearts. J Cardiovasc Pharmacol 19:929–93
54. Yanagisawa M, Kurihara H, Kimura S (1988) A novel potent vasoconstrictor peptide produced by vascular endothelial cells. Nature 332:411–415
55. Yang Z, Richard V, Von Segesser L, Bauer E, Stulz P, Turina M, Lüscher TF (1990) Threshold concentrations of endothelin-1 potentiate contractions to norepinephrine and serotonin in human arteries: a new mechanism of vasospasm? Circulation 82:188–195
56. Yasuda M, Kohno M, Tahara A, Itagane H, Toda I, Akioka K, Teragaki M, Oku H, Takeuchi K, Takeda T (1990) Circulating immunoreactive endothelin in ischemic heart disease. Am Heart J 119:801–806
57. Yorikane R, Koike H (1990) The arrhythmogenic action of endothelin in rats. Jpn J Pharmacol 53:259–263

Endothelin and Renal Function: From Theory to Therapy

T. J. Rabelink and P. Morrison

Introduction

The discovery of endothelin has opened up a new proliferative field in biomedical research. The kidney appears to be a particular important target organ for the effects of endothelins, and endothelins have been incriminated in the pathogenesis of several diseases of the kidney, including renal injury following ischaemia (Mino et al. 1992; Brooks et al. 1994) and septicemia (Badr 1992), administration of cyclosporin (Kon et al. 1990) and radiocontrast (Margulies et al. 1992; Heyman et al. 1992), glomerulonephritis (Simonson and Dunn 1991), and hepatorenal syndrome (Moore et al. 1992). Although experimental studies suggest an important role for endothelin as a pathogenetic factor in renal disease, proof in humans is still lacking. However, the introduction of endothelin-receptor antagonist in clinical medicine within the very near future will create opportunities to evaluate its role in renal disease. This review will briefly summarize the renal endothelin system and its possible involvement in renal diseases. Subsequently, we will discuss whether these concepts may be translated into therapies for clinical renal medicine.

Renal Endothelin Production

Three distinct endogenous endothelin (ET) isopeptides, ET-1, ET-2 and ET-3, have been recognized, which are all cleaved from proendothelins (big ETs) by an endothelin-converting-enzyme (Yanagisawa et al. 1988; Masaki et al. 1991). ET-2 seems to be produced exclusively by the intestine, and therefore will not be discussed (Saida et al. 1989). The kidney is an important organ for endothelin synthesis, and ET-1 is produced by endothelial, mesangial, glomerular epithelial and medullary collecting duct cells (Kohan 1991, 1992, 1993; Karet et al. 1993a). Although studies in rats and cell cultures also demonstrated ET-3 production in mesangial cells and glomerular epithelial cells, no ET-3 production in the human kidney could be demonstrated (Karet et al. 1993a). These data suggest that ET-1 is the clinically most important endothelin in human renal medicine. The regulation of endothelial ET-1 synthesis has been discussed in detail before and includes thrombin, hypoxia and shear stress (Lüscher, this volume). Mesangial ET-1 synthesis is stimulated by various humeral factors involved in glomerular injury, such as thrombin, transforming growth factor β and angiotensin II (Sakamoto et al. 1992). In contrast, atrial natriuretic peptide and nitric oxide inhibit mesangial ET-1 synthesis (Sakamoto et al. 1992). The regulation of ET synthesis by medullary cells is still largely unknown, but may include factors such as osmolarity (Kohan 1993). Although ET may be released into the vascular and the interstitial compartment, it is believed that endothelial produced ET is predominantly abluminally secreted (Lüscher, this volume). The polarity of ET-release from renal epithelial and mesangial cells is not presently known.

Endothelin Receptors and Signalling in the Kidney

The initial classification of the endothelin receptors was based on the comparative agonist-activity of the endothelins. In this way an ET_A receptor, which has a greater affinity for ET-1 than ET-3, and an ET_B receptor, which has equal affinity for the endothelin isopeptides were proposed. The presence of these two different receptors was subsequently confirmed by cloning (Arai et al. 1990; Sakurai et al. 1990). Recent studies in human kidneys, using receptor autoradiograpy, have shown that the ET_B receptor subtype predominates over the ET_A receptor (Nambi et al. 1992; Karet et al. 1993b). The ET_A receptor is localized to the vasculature, while ET_B distribution is more widespread. Interestingly, ET receptor expression and ET synthetic sites show parallel distribution in the kidney, suggesting an autocrine/paracrine mode of action (Marsen et al. 1994).

After binding to the ET_A receptor ET-1 activates phospholipase C via a pertussistoxin insensitive G-protein, resulting a. o. in the formation of inositol triphosphate (IP3) (for review see Rubanyi and Polokoff 1994). This IP3 formation will stimulate intracellular calcium mobilization, which activates L-type calcium channels, thus contributing to sustained increments in intracellular calcium. ET-receptor activation perhaps may also directly open these channels (Goto et al. 1989). It has been postulated that these signal transduction pathways are mainly involved in the contractile actions of ET-1. Phospholipase C activation will also stimulate protein kinase C (PKC), and several protein tyrosine kinases (PTK) (Simonson 1994). These cytosolic effectors in turn alter gene expression, and have been implicated in the mitogenic actions of ET-1, while PKC may perhaps also contribute to vasoconstriction by phosphorylation of myosin light chains. ET-receptor activation can also stimulate arachidonic metabolism via activation of phospholipase A2, which may result in the formation of both vasodilating prostaglandins (e.g. PGI2) as well as constricting prostaglandins (TXA2) in the kidney (De Nucci 1988, Rubanyi and Polokoff 1994). Finally, recent studies have indicated that ATP-sensitive postassium channels may also be involved in the signalling of ET-1 (Haynes and Webb 1991). However, their relevance for the kidney has not been established yet.

Physiological Actions of Endothelin in the Kidney

ET-1 may act both in an endocrine (ET-1 derived from the systemic circulation) as well as an autocrine/paracrine signalling mode (ET-1 produced in the kidney). Infusions of ET-1 probably mimic the former mode of action, while the autocrine/paracrine actions can only be evaluated by studying tissue expression of endothelin and the effects of intrarenal administration of endothelin antagonists.

Administration of low dosages of ET-1 in humans that increase plasma levels two-fold produce long-lasting increments renal vasoconstriction (Weitzberg et al. 1991; Rabelink et al. 1994; Sorensen et al. 1994). In vitro studies have shown that this reduction in renal blood flow is caused by constriction of the glomerular afferent and efferent arterioles, and the arcuate and interlobar arteries (Badr et al. 1989a; King et al. 1989; Kon et al. 1989; Edwards et al. 1990). In addition, ET-1 can reduce ultrafiltration coefficient by mesangial contraction (Badr et al. 1989b). Controversy exists with regard to the receptor subtypes mediating these renal hemodynamic effects. The distribution and the role of the receptor subtypes varies between tissues and between species. This is of particular relevance for the kidney. For example, in the rat the renal vascular effects of endothelin can be completely blocked by a non selective ET_A/ET_B receptor blocker, but not by ET_A selective blockers (Warner et al. 1993, Wellings et al. 1993), while the ET_B agonist sarafatoxin 6c mimics these renal vasoconstrictive effects (Gellai et al. 1994). This suggests that the hemodynamic effects of endothelin in the rat are predominantly mediated by the ET_B receptor. In contrast, endothelin-induced renal vasoconstriction in dogs and rabbits is completely blocked by ET_A receptor antagonists (Brooks et al. 1994; Télémaque et al. 1993) and ET_B agonists caused vasodilation rather than vasoconstriction (Yakimura et al. 1994). The localization of the ET_A receptor to the vasculature in human kidneys suggests that this receptor subtype could also be primarily involved in the hemodynamic effects of endothelin in man (Karet et al. 1993b).

Administration of ET-1 in humans also consistently induces sodium retention (Rabelink et al. 1994; Sorensen et al. 1994; Bijlsma et al. 1995; Kaasjager et al. 1995), even during infusion of very low dosages of ET-1 that do not affect renal blood flow, GFR or renin angiotensin system (Rabelink et al. 1994). In contrast, studies in the isolated perfused rat kidney have consistently shown a natriuretic effect of ET-1 (Ferrario et al. 1989; Perico et al. 1990a). Similarly, several rat studies report a natriuretic effect of low dose systemic administration of ET-1, despite a fall in GFR and renal blood flow (Perico et al. 1991; Harris et al. 1991). This natriuresis may be due to stimulation of ANP (Munger et al. 1991) or pressure-natriuresis secondary to a higher blood pressure during endothelin infusion (King et al. 1989). In addition, evidence is now accumulating that endothelin can directly inhibit Na/K ATPase activity in rat inner medullary collecting duct cells (Zeidel et al. 1989). ET-1 can also inhibit the effects of vasopressin in these cells (Oishi et al. 1991). These data indicate that the net effect of ET-1 on sodium excretion depends on a balance between sodium retaining and natriuretic factors. In humans the sodium retaining effects of ET-1 predominate with only small increases in plasma ET-1.

The physiological importance of possible autocrine actions of ET-1 in the kidney has not been defined yet. Studies using intrarenal administration of ET receptor blockers showed no or little effect on renal blood flow and sodium excretion (Pollock and Opgenoth 1994; Davis et al. 1994; Télémaque et al. 1993) under basal conditions while one study demonstrated an increase in renal blood flow and sodium excretion (Brooks et al. 1994).

Endothelin as a Pathogenetic Factor in Kidney Disease

At low concentrations the effects of ET-1 on glomerular microcirculation may help to maintain perfusion pressure. However, the sustained reductions of GFR and renal blood flow observed during administration of ET-1 in humans may endanger oxygen supply within the kidney (Oken 1983). Such effects may contribute to renal injury in clinical conditions characterized by elevated plasma levels of ET. Elevated plasma levels of ET-1 have been reported in many renal diseases, including cyclosporin (Fogo et al. 1990) and radiocontrast administration (Margulies et al. 1992), acute renal failure (Tomita et al. 1989) and chronic renal failure (Koyama et al. 1989). This may be due to impaired ET-1 clearance by neutral endopeptidases in the brush border of the proximal tubule (Abassi et al. 1992). In addition, renal ET-1 production is enhanced in several models of experimental renal diseases, such as ischemia and administration of cyclosporin and radiocontrast (Roubert et al. 1993; Heyman et al. 1992; Nakahama 1990). Since, renal vasoconstriction is a hallmark of these nephropathies, it seems logical to assume an important role for endothelin in the pathogenesis of these conditions. In agreement with such a role intrarenal administration of ET-antiserum in rats could reverse renal vasoconstriction following ischemia and administration of cyclosporin and endotoxin (Kon et al. 1989; Kon and Badr 1991 ; Morise et al. 1994). Treatment with a monoclonal antibody to ET could also protect from histological damage, and attenuate increments in serum creatinine following renal ischemia (Shibouta et al. 1990).

Such protective effects of ET-antibodies may not only be caused by antagonism of ET-induced glomerular vasoconstriction. ET-1 may also have effects on cell proliferation. ET-1 has been demonstrated to stimulate growth in cultures of (human) mesangial cells (Bakris and Re 1993; Simonson et al. 1989). In addition, it can potentiate the actions of other growth factors and it increases the expression of extracellular matrix proteins (Funabiki et al. 1990; Simonson 1994). Both compensatory renal growth and matrix expansion are characteristics of progressive renal disease. In rats with renal mass ablation ET-1 expression in the kidney was markedly enhanced and increased in time parallel to the development of glomerulosclerosis (Benigni et al. 1991; Orisio et al. 1993), suggesting that ET-1 may also play a role in the progression of renal failure by its actions as a growth factor.

Endothelin as a Therapeutic Target

These experimental studies suggest that endothelin may play a role in the pathogenesis of several renal diseases. The introduction of endothelin receptor blockers will allow clinicians to test the potential of endothelin as a therapeutic target in human renal disease. Several endothelin receptor antagonists have now been identified. The first and most widely used one is BQ123, a cyclic pentapeptide, which selectively inhibits the ET_A receptor (Ihara et al. 1992). Subsequently several peptide analogues were developed, which could block both the ET_A and the ET_B receptor. However, a major problem with these peptide-analogues is that they are not effective via the oral route. Moreover, binding of endothelin to the receptor is almost irreversible, making dis-

Table 1. Endothelin receptor blockers most frequently used

Name	Supplier	Class	Selectivity	Administration	Reference
Peptide analogues					
BQ-123	Banyu	Cyclic pentapeptide	ET_A	i.v.	Ihara et al. (1992)
BQ-788	Banyu	Cyclic pentapeptide	ET_B	i.v.	Mase et al. (1993); Fukuroda et al. (1994)
RES-701-1	Kyowa Hakko Kogyo	Cyclic pentapeptide	ET_B	i.v.	Karaki et al. (1994); Tanaka et al. (1994)
TAK-044	Takedu	Cyclic pentapeptide	ET_A/ET_B	i.v.	Kusumoto et al. (1994)
FR-139 317	Fujisawa	Acyl tripeptide	ET_A	i.v.	Sogabe et al. (1993)
PD 145 065	Parke-Davis	Hexapeptide	ET_A/ET_B	i.v.	Cody et al. (1993)
IRL 1038	Ciba-Geigy	Hexapeptide	ET_B	i.v.	Karaki et al. (1993)
Non-peptides					
Ro 47-0203 (bosentan)	Hoffmann-La Roche	Benzene-sulfonamide	ET_A/ET_B	oral/i.v.	Clozel et al. (1994); Teerlink et al. (1994)
BMS-182 874	Bristol-Meyer Squibb	Naphthalene-sulfonamide	ET_A	oral	Stein et al. (1994)
SB 209 670	Smith Kline Beecham	Indanecar-boxylic acid	ET_A/ET_B	i.v.	Ohlstein et al. (1994)

placement by receptor antagonists more difficult (Wu-Wong et al. 1994). This explains the high dosages required in experimental studies. These characteristics and the high costs of the peptide analogues render these receptor-antagonists less suitable for use in humans. Recently, non-peptide ET-receptor antagonists are being introduced, some of which are also orally active (see Table 1). However, the safety and applicability of these compounds in humans is still unclear. In particular it is unclear whether the ET_B receptor plays a role in renal vasoconstriction or injury in humans. Most of the orally active blockers entering clinical medicine probably will be mixed ET_A/ET_B receptor blockers. Although it may seem a practical approach to block both receptors, it could also implicate (undesirable) blockade of ET_B mediated autocrine actions in the kidney such as e.g. vasodilation or inhibition of sodium excretion. Moreover, recent studies linked familial Hirschsprungs disease to a defect in the intestinal ET_B receptor (Puffenberg et al. 1994).

Endothelin receptor blockers have been applied in several models of experimental renal disease.

Cyclosporin Nephropathy. Intrarenal administration of the ET_A receptor blocker BQ123 could prevent cyclosporin-induced vasoconstriction (Kon et al. 1990). However, the data on the effectiveness of *systemic* administration of endothelin-receptor blockers in rats vary from no effect (Fogo et al. 1992; Davis et al. 1994) to antagonist action

(Kivlighn et al. 1994; Perico et al. 1990b). It is important to point out that these studies were all done in rats, while the distribution of ET-receptors may vary with species. Moreover, these studies focused on antagonism of glomerular hemodynamic effects, while cyclosporin nephrotoxicity is more likely to be determined by tubular and interstitial injury.

Radiocontrast Nephropathy. After administration of radiocontrast compensatory formation of prostaglandins probably counteracts iodine-induced decrements in renal blood flow (Parvez et al. 1988). However, during prostaglandin synthesis inhibition, prolonged iodine-induced vasoconstriction is observed, which could be completely reversed by endothelin receptor blockade (Cantley et al. 1992; Oldroyd et al. 1994). One thus could postulate that endothelin blockers may be of use in preventing radiocontrast nephropathy in patients with dysfunctional renal prostaglandin synthesis such as in volume depletion, renal failure and in heart failure, who are known to run a higher risk of developing contrast nephropathy.

Ischemia. Recently, both ET_A as well as the ET_A/ET_B receptor blockade has been shown to attenuate the sustained reductions in renal perfusion after ischemia in rats (Mino et al. 1992; Clozel et al. 1993) (Fig. 1). However, in other species such as the dog BQ123 had no effect on the reduction in the glomerular filtration rate (GFR) following renal ischemia, despite the fact that renal vasoconstriction was attenuated (Brooks et al. 1994), again underscoring the possible influence of species differences and the contribution of other (tubular) processes. Importantly, these studies were all done during local administration of the endothelin-blockers. It should be noticed that endothelin may also have a role in maintenance of vascular tone (Haynes and Webb 1994). This may be particularly relevant in conditions that lead to renal ischemia in vivo. Studies in rats with hemorrhagic shock showed that endothelin blockade can cause sustained reductions in blood pressure (Zimmerman et al. 1994). It is conceivable that such changes during systemic blockade of endothelin may have an adverse

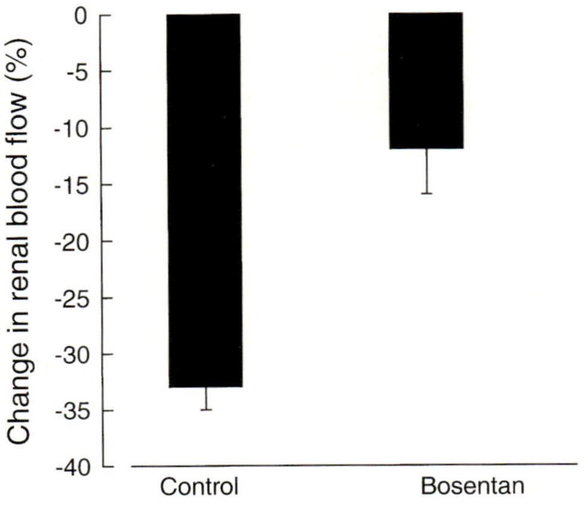

Fig. 1. Effects of intrarenal administration of the mixed ET_A/ET_B receptor blocker bosentan on the sustained reduction in renal blood flow following 45 min clamping of the renal artery. [Adapted from Clozel et al. (1993)]

effect on recovery of ischemic renal failure. Such systemic effects of endothelin-blockade may also be relevant for acute renal failure due to endotoxemia.

Chronic Renal Failure. Finally, chronic administration of a selective ET_A receptor blocker, as well as an ET_A/ET_B blocker to rats with remnant kidneys prevented the development of proteinuria and glomerulosclerosis (Benigni et al. 1993, 1994). There was no difference between the two blockers in their effects on progressive renal disease, suggesting that the ET_B receptor is less important in this process in rats. However, in these studies the ET-receptor blockers also decreased blood pressure, which may have slowed the progression of glomerulosclerosis independent of effects on the mitogenic actions of endothelin.

Although these experimental data suggest a therapeutic effect of ET-receptor blockers in renal disease, the experimental design of these studies does not allow translation into clinical applicability yet.

The clinician can also try to block the effects of ET-1 by interfering with calcium-dependent signal transduction. As mentioned before, dihydropyridine sensitive channels probably play an important role in the signal transduction following endothelin-receptor activation (Rubanyi and Polokoff 1994). This would imply that calcium antagonists may interfere with these processes. However, the effectiveness of calcium channel blockade to interfere with the constrictor actions of ET-1 seems to depend on the vascular bed studied (Lüscher, this volume). Controversy also exists regarding the interaction of calcium channel blockade and the renal effects of endothelin. Again, depending on the species studied and the dosages of endothelin and calcium blocker used, no effect as well as near-total reversal vasoconstriction has been observed (Cao and Banks 1990; Hof et al. 1990; Takahashi et al. 1993). Recent studies in humans demonstrate that normal oral maintenance dosages of nifedipine attenuated the constrictor response and antinatriuresis following systemic endothelin infusion in humans (Kaasjager et al. 1995). These effects are probably specific for calcium channel blockers since renal vasodilation with the ACE inhibitor enalapril could not attenuate ET-1 induced vasoconstriction and sodium retention (Fig. 2). Calcium channel blockers clearly interfere only with part of the constrictor actions of endothelin, and mitogenic effects may be independent of calcium influx. However, calcium channel blockers may also inhibit other mechanisms that can cause renal injury, such as cellular calcium overload, mitochondrial dysfunction and ischemia/reperfusion (Arnold et al. 1985; Burke et al. 1984). These are injury mechanisms that also occur in many clinical conditions characterized by elevated plasma levels of endothelin. Indeed, calcium antagonist have been shown to be able to prevent renal injury in various experimental models of acute renal failure such as glycerol, gentamycin, radiocontrast and endotoxin administration, and ischemia (Schrier et al. 1987). Importantly, there are also indications that calcium antagonists can prevent cyclosporin and radiocontrast induced renal vasoconstriction and reductions in GFR in humans (Neumayer et al. 1989; Wagner et al. 1987).

Other ways to interfere with the endothelin system probably will emerge in the near future as well. These include inhibition of ATP sensitive potassium channels (Haynes and Webb 1991) and the development of endothelin-converting enzyme

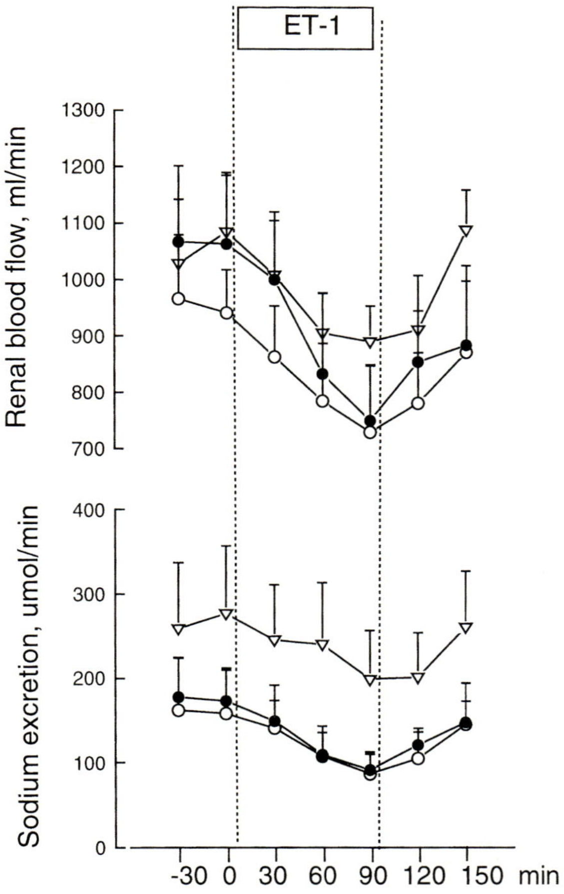

Fig. 2. Effects of infusion of ET-1 (2.5 ng kg-1 min-1), yielding pathophysiological increments in plasma ET-1, on renal blood flow (*top*), and sodium excretion (*bottom*) in healthy volunteers during under control conditions (*open circles*), after 5 days pretreatment with 20 mg enalapril b.i.d. (*closed circles*) and after 5 days treatment with 60 mg nifedipine GITS o.d. (*triangles*). The decrease in renal blood flow was larger during enalapril ($p < 0.05$), and less during nifedipine ($p < 0.05$) compared to control. Sodium excretion during nifedipine was significantly elevated throughout the experiment compared to control or enalapril ($p < 0.05$). [Adapted from Kaasjager et al. (1995)]

inhibitors. The latter may be of particular relevance as in humans the kidney probably is the most important organ converting circulating big-ET-1 into ET-1 (Ahlborg et al. 1994).

Conclusions

Experimental studies have shown that the endothelin system is activated in renal disease models and that blockade of the effects of endothelin may reverse renal impairment. However, these studies cannot be easily extrapolated to the human situation. There are marked interspecies differences with regard to ET-receptor distribution. Moreover, most of these studies used intrarenal administration of blockers, and the side effects of systemic blockade of ET-receptors have not been identified. Finally, studies thus far focused on antagonism of the hemodynamic actions of ET-1, while more hard endpoints such as histological damage are lacking.

Therefore, at this stage, specific endothelin antagonist are exciting research tools, which will help to understand renal function and disease. However, studies in humans or human disease models are needed to assess the clinical future of these drugs. In the meantime the clinician may consider to use calcium antagonists as an alternative way to block the actions of ET-1. Clearly, calcium antagonists inhibit only part of the actions of endothelin. On the other hand it is unlikely that endothelin is the only pathogenetic mechanism in renal disease, and calcium antagonists interfere with other injury mediators as well. Moreover, there is extensive experience on the safety and efficacy of these drugs in renal disease.

Acknowledgement. Ton Rabelink is sponsored by a fellowship of the Royal Dutch Academy of Sciences.

References

Abassi ZA, Tate JE, Golomb E, Keiser HR (1992) Role of neutral endopeptidase in the metabolism of endothelin. Hypertension 20:89–95

Ahlborg G, Ottosson-Seeberger A, Hemsén A, Lundberg J (1994) Big ET-1 infusion in man causes renal ET-1 release, renal and splanchnic vasoconstriction, and increased mean arterial blood pressure. Cardiovasc Res 28:1559–1563

Arai H, Hori S, Aramori I, Ohkubo H, Nakanishi S (1990) Cloning and expression of a cDNA encoding an endothelin receptor. Nature 348:730–732

Arnold PE, Lumlertgul D, Burke TJ, Schrier RW (1985) In vitro versus in vivo mitochondrial calcium loading in ischemic acute renal failure. Am J Physiol 248:F845–F850

Badr KF (1992) Sepsis-associated renal vasoconstriction: potential targets for future therapy. Am J Kidney Dis 20:207–213

Badr KF, Murray JJ, Breyer MD, Takahashi K, Inagami T, Harris RC (1989a) Mesangial cell, glomerular and renal vascular responses to endothelin in the rat kidney. J Clin Invest 83:336–342

Badr KF, Murray JJ, Breyer MD, Takahashi K, Inagami T, Harris RC (1989b) Mesangial cell, glomerular and renal vascular responses to endothelin in the rat kidney: elucidation of signal transduction pathways. J Clin Invest 83:336–342

Bakris GL, Re RN (1993) Endothelin modulates angiotensin II-induced mitogenesis of human mesangial cells. Am J Physiol 264:F937–F942

Benigni A, Perico N, Gaspari F, Zoja C, Bellizzi L, Gabanelli M, Remuzzi G (1991) Increased renal endothelin production in rats with reduced renal mass. Am J Physiol 260:F331–F339

Benigni A, Zoja C, Corna D, Orisio S, Longaretti L, Bertani T, Remuzzi G (1993) A specific endothelin subtype A receptor antagonist protects against injury in renal disease progression. Kidney Int 44: 440–444

Benigni A, Zoja C, Corna D, Orisio S, Remuzzi G (1994) Role of renal endothelin (ET) type B receptor (R) in experimental progressive renal disease. J Am Soc Nephrol 5:573 (abstract)

Bijlsma JA, Rabelink TJ, Kaasjager KAH, Koomans HA (1995) L-Arginine does not prevent the renal effects of endothelin in humans. J Am Soc Nephrol 5:1508–1516

Brooks DP, DePalma PD, Gellai M, Nambi P, Ohlstein EH, Elliot JD, Gleason JG, Ruffolo RR (1994) Nonpeptide endothelin receptor antagonists. III. Effect of SB 209670 and BQ 123 on acute renal failure in anesthetized dogs. J Pharmacol Exp Ther 271:769–775

Burke TJ, Arnold PE, Gordon JA, Bulger RE, Dobyan DC, Schrier RW (1984) Protective effect of intrarenal calcium membrane blockers before or after renal ischemia. Functional, morphological and mitochondrial studies. J Clin Invest 74:1830–1841

Cantley LG, Spokes K, Clark B, Kulik A, Epstein FH (1992) An endothelin receptor antagonist decreases the effect of iothalamate on renal blood flow. J Am Soc Nephrol 3:434–437

Cao L, Banks RO (1990) Cardiovascular and renal actions of endothelin: effects of calcium channel blockers. Am J Physiol 258: F254–F258

Clozel M, Breu V, Burri K, Cassal JM, Fischli W, Gray G, Hirth G, Loffler BM, Muller M, Neidhart W, Ramuz H. (1993). Pathophysiological role of endothelin revealed by the first orally active endothelin receptor antagonist. Nature 365: 759–761

Clozel M, Breu V, Gray GA, Kalina B, Loeffler B, Burri K, Cassal J, Hirth G, Mueller M, Neidhart W, Ramuz H (1994) Pharmacological characterization of bosentan, a new potent orally active nonpeptide endothelin receptor antagonist. J Pharmacol Exp Ther 270: 228–235

Cody WL, Doherty AM, He JX, Depue PL, Waite LA, Topliss JG, Haleen SJ, Ladouceur KE, Flynn MA, Hill DM, Reynolds EE (1993) The rational design of a highly potent combined ET_A and ET_B receptor antagonist (PD 145 065) and related analogues. Med Chem Res 3: 154–162

Davis LS, Haleen SJ, Doherty AM, Cody WL, Keiser JA (1994) Effects of selective endothelin antagonists on the hemodynamic response to cyclosporin A. J Am Soc Nephrol 4: 1448–1454

De Nucci G, Thomas R, D'Orleans-Juste P, Antunes E, Walder C, Warner TD, Vane JR (1988) Pressor effects of circulating endothelin are limited by its removal in the pulmonary circulation and by the release of prostacyclin and endothelium-derived relaxing factor. Proc Natl Acad Sci USA 85: 9797–9800

Edwards RM, Trizna W, Ohlstein EH (1990) Renal mivrovascular effects of endothelin. Am J Physiol 259: F217–F221

Ferrario RG, Foulkes R, Salvati P, Patrono C (1989) Hemodynamic and tubular effects of endothelin and thromboxane in the isolated perfused rat kidney. Eur J Pharmacol 171: 127–134

Fogo A, Hakim RC, Sugiara M, Inagami T, Kon V (1990) Severe endothelial injury in a renal transplant patient receiving cyclosporine. Transplantation 49: 1190–1192

Fogo A, Hellings S, Inagami T, Kon V (1992) Endothelin receptor anatagonism is protective in in vivo acute cyclosporine toxicity. Kidney Int 42: 770–774

Fukuroda T, Fujikawa T, Ozaki S, Ishikawa K, Yano M, Nishikibe M (1994) Clearance of circulating endothelin-1 by ET_B receptors in rats. Biochem Biophys Res Commun 199: 1461–1465

Funabiki KJ, Horikoshi S, Tomino Y, Nagai Y, Koide H (1990) Immunohistochemical analysis of extracellular components in the glomerular sclerosis of patients with glomerulonephritis. Clin Nephrol 34: 239–246

Gellai M, DeWolf R, Pullen M, Nambi P (1994) Distribution and functional role of renal ET receptor subtypes in normotensive and hypertensive rats. Kidney Int 46: 1287–1294

Goto K, Kasuya Y, Matsuki N, Takuwa Y, Kurihara H, Ishikawa T, Kimura S, Yanagisawa M, Masaki T (1989) Endothelin activates the dihydropyridine-sensitive, voltage-dependent Ca^{2+} channel in vascular smooth muscle. Proc Natl Acad Sci USA 86: 3915–3918

Harris PJ, Zhuo J, Mendelsohn FAO, Skinner SL (1991) Haemodynamic and renal tubular effects of low doses of endothelin in anesthetized rats. J Physiol Lond 433: 25–39

Haynes WG, Webb DJ (1991) Selective antagonism of endothelin-1 induced vasoconstriction by the potassium channel opening agent cromakalim. J Hypertens 9: 1089

Haynes WG, Webb DJ (1994) Contribution of endogenous generation of endothelin-1 to basal vascular tone. Lancet 344: 852–854

Heyman SN, Clark BA, Kaiser N, Spokes K, Rosen S, Brezis M, Epstein FM (1992) Radiocontrast agents induce endothelin release in vivo and in vitro. J Am Soc Nephrol 3: 58–62

Hof RP, Hof A, Takiguchi T (1990) Attenuation of endothelin-induced regional vasoconstriction by isradipine: a nonspecific antivasoconstrictor effect. J Cardiovasc Pharmacol 15: S48–S54

Ihara M, Noguchi K, Saeki T, Fukuroda T, Tsuchida S, Kimura S, Fukami T, Ishikawa K, Nishikibe M, Yano M (1992) Biological profiles of highly potent endothelin antagonists selective for the ET_A receptor. Life Sci 50: 247–251

Kaasjager KAH, Koomans HA, Rabelink TJ (1995) Effectiveness of enalapril versus nifedipine to antagonize blood pressure and renal response to endothelin in humans. Hypertension (in press)

Karaki H, Sudjarwo SA, Hori M, Sakata K, Urade Y, Takai M, Okada T (1993) ET_B receptor antagonist, IRL 1038, selectively inhibits the endothelin-induced endothelium-dependent vascular relaxation. Eur J Pharmacol 231: 371–374

Karaki H, Sudjarwo SA, Hori M, Tanaka K, Matsuda Y (1994) Effects of a novel endothelin ET_B receptor antagonist, RES-701-1, on isolated blood vessels. Jpn J Pharmacol 64 [Suppl 1]:167P

Karet F, Plumpton C, Davenport A (1993a) Human kidney: endothelin isoforms revealed by HPLC with RIA, and receptor subtypes detected using ligands BQ123 and BQ30320. J Cardiovasc Pharmacol 22:29–33

Karet F, Kuc R, Davenport A (1993b) Novel ligands BQ123 and BQ3020 characterise endothelin receptor subtypes ET_A and ET_B in human kidney. Kidney Int 44 (1):36–42

King AJ, Brenner BM, Anderson S (1989) Endothelin: a potent renal and systemic vasoconstrictor peptide. Am J Physiol 256:F1051–F1058

Kivlighn SD, Gabel RA, Siegl PKS (1994) Effects of BQ-123 on renal function and acute

cyclosporine-induced renal function. Kidney Int 45:131–136

Kohan DE (1991) Endothelin synthesis by renal tubule cells. Am J Physiol 261:F221–F226

Kohan DE (1992) Production of endothelin-1 by rat mesangial cells: regulation by tumor necrosis factor. J Lab Clin Med 119:477–484

Kohan DE (1993) Endothelin production by human inner medullary collecting ducts. J Am Soc Nephrol 3:1719–1721

Kohan DE, Padilla E (1993) Osmolar regulation of endothelin-1 production by rat inner medullary collecting duct. J Clin Invest 91:1235–1240

Kon V, Badr KF (1991) Biological actions and pathophysiological significance of endothelin in the kidney. Kidney Int 40:1–12

Kon V, Yoshioka T, Fogo A, Ichikawa I (1989) Glomerular actions of endothelin in vivo. J Clin Invest 83:1762–1767

Kon V, Sugiura M, Inagami T, Harviue BR, Ichikawa I, Hoover RL (1990) Role of endothelin in cyclosporine induced glomerular dysfunction. Kidney Int 37:1487–1491

Koyama H, Nishzawa Y, Morii NH, Tabate T, Inoue T, Yamaji T (1989) Plasma endothelin levels in patients with uraemia. Lancet 1:991–992

Kusumoto K, Kubo K, Kandori H, Kitayoshi T, Sato S, Wakimasu M, Watanabe T, Fujino M (1994) Effects of a new endothelin antagonist, TAK-044, on post-ischemic acute renal failure in rats. Life Sci 55:301–310

Margulies KB, McKinley LJ, Burnett JC (1992) Endothelin in human and canine radiocontrast-induced nephropathy. J Vasc Res 29:163–164

Marsen TA, Schramek H, Dunn MJ (1994) Renal activity of endothelin: linking cellular signaling pathways to disease. Kidney Int 45:336–344

Masaki T, Kimura S, Yanagisawa M, Goto K (1991) Molecular and cellular mechanism of endothelin regulation. Circulation 84:1457–1468

Mase T, Fukami T, Nagase T, Yamakawa T, Takahashi H, Naya A (1993) 14th medicinal chemical symposium, Shizuoka, abstract P28:59

Mino N, Kobayashi M, Nakajima A, Amano H, Shimamoto, Ishikawa K, Watanabe K, Nishikibe M, Yano M, Ikemoto F (1992) Protective effect of a selective endothelin receptor antagonist, BQ-123, in ischaemic acute renal failure in rats. Eur J Pharmacol 221:77–81

Moore K, Wendon J, Frazer M, Karani J, Williams R, Badr K (1992) Plasma endothelin immunoreactivity in liver disease and hepatorenal syndrome. N Engl J Med 327:1774–1778

Morise Z, Ueda M, Aiura K, Endo M, Kitajima M (1994) Pathophysiologic role of endothelin-1 in renal function in rats with endotoxin shock. Surgery 115:199–204

Munger KA, Sugiura M, Takahashi K, Inagami T, Badr KF (1991) A role for atrial natriuretic peptide in endothelin-induced natriuresis. J Am Soc Nephrol 1:1278–1283

Nakahama H (1990) Stimulatory effect of cyclosporine A on endothelin secretion by a cultured renal epithelial cell line, LLC-PK1 cells. Eur J Pharmacol 180:191–192

Nambi P, Pullen M, Wu HL, Aiyar N, Ohlstein EH, Edwards RM (1992) Identification of endothelin receptor subtypes in human renal cortex and medulla using subtype-selective ligands. Endocrinology 131:1081–1086

Neumayer HH, Junge W, Kufner A, Wenning A (1989) Prevention of radiocontrast-media-induced nephrotoxicity by the calcium channel blocker nitrendipine: a prospective randomized clinical trial. Nephrol Dial Transplant 4:1030–1036

Ohlstein EH, Beck JR GR, Douglas SA, Nambi P, Lago MA, Gleason JG, Ruffolo JR RR,

Feuerstein G, Elliot JD (1994) Nonpeptide endothelin receptor antagonists. II. Pharmacological characterization of SB 209 670. J Pharmacol Exp Ther 271:762–768

Oishi R, Nonoguchi H, Tomita K, Marumo F (1991) Endothelin-1 inhibits AVP-stimulated osmotic water permeability in rat medullary inner collecting duct. Am J Physiol 261: F951–F956
Oken DE (1983) Theoretical analysis of pathogenetic mechanisms in experimental acute renal failure. Kidney Int 24: 16–26
Oldroyd S, Slee SJ, Haylor J, Morcos K, Wilson C (1994) Role for endothelin in the renal responses to radiocontrast media in the rat. Clin Sci 87: 427–434
Orisio S, Benigni A, Bruzzi I, Corna D, Perico N, Zoja C, Benatti L, Remuzzi G (1993) Renal endothelin gene expression is increased in remnant kidney and correlates with disease progression. Kidney Int 43: 354–358
Parvez Z, Marsan RE, Moncada R, Patel N (1988) Effect of contrast media on prostaglandin synthesis in vivo. Invest Radiol 23 [Suppl]: S178–S181
Perico N, Dadan J, Cabanelli M, Remuzzi G (1990a) Cyclooxygenase products and atrial natriuretic peptide modulate renal response to endothelin. J Pharmacol Exp Ther 252: 1213–1220
Perico N, Dadan J, Remuzzi G (1990b) Endothelin mediates the renal vasoconstriction induced by cyclosporin in the rat. J Am Soc Nephrol 1: 76–83
Perico N, Cornejo RP, Benigni A, Malanchine B, Ladny JR, Remuzzi G (1991) Endothelin induces diuresis and natriuresis in the rat by acting on proximal tubular cells through a mechanism mediated by lipooxygenase products. J Am Soc Nephrol 2: 57–69
Pollock DM, Opgenorth TJ (1994) ET_A eceptor-mediated responses to endothelin-1 and big endothelin-1 in the rat kidney. Br J Pharmacol 11: 729–732
Puffenberger EC, Hosoda K, Washington SS, Nakao K, de Wit D, Yanagisawa M, Chakravasti A (1994) A missense mutation of the endothelin B receptor gene in multigenic Hirschsprung's disease. Cell 79: 1257–1267
Rabelink TJ, Kaasjager KAH, Boer P, Stroes EG, Braam B, Koomans HA (1994) Effects of endothelin-1 on renal function in humans: implications for physiology and pathophysiology. Kidney Int 46: 376–381
Roubert P, Gillard-Roubert V, Pourmarin L, Cornet S, Guilmard C, Plas P, Pirotzky E, Chabrier PE, Braquet P (1993) Endothelin receptor subtypes A and B are up-regualted in an experimental model of acute renal failure. Am Soc Pharmacol Exp Ther 45: 182–188
Rubanyi GM, Polokoff MA (1994) Endothelins: molecular biology, biochemistry, pharmacology, physiology, and pathophysiology. Pharmacol Rev 46: 325–415
Saida K, Mitsui Y, Ishida N (1989) A novel peptide, vasoactive intestinal contractor, of a new (endothelin) peptide family. J Biol Chem 264: 14613–14616
Sakamoto H, Sasaki S, Nakamura Y, Fushimi K, Marumo F (1992) Regulation of endothelin-1 production in cultured rat mesangial cells. Kidney Int 41: 350–355
Sakurai T, Yanagisawa M, Takuwa Y, Miyazaki H, Kimura S, Goto K, Masaki T (1990) Cloning of a cDNA encoding a non-isopeptide-selective subtype of the endothelin receptor. Nature 348: 732–735
Schrier RW, Arnold PE, Van Putten VJ, Burke TJ (1987) Cellular calcium in ischemic acute renal failure: role of calcium entry blockers. Kidney Int 32: 313–321
Shibouta Y, Suzuki N, Shino A, Matsumuto H, Terashita Z-I, Kondo K, Nishikawa K (1990) Pathophysiological role of endothelin in acute renal failure. Life Sci 46: 1611–1618
Simonson MS (1994) Endothelin peptides and compensatory growth of renal cells. Curr Opin Nephrol Hypertens 3: 73–85
Simonson MS, Dunn MJ (1991) Endothelin peptides: a possible role in glomerular inflammation. Lab Invest 64: 1–4
Simonson MS, Wann S, Mene P (1989) Endothelin stimulates phospholipase C, Na^+/H^+ exchange, *c-fos* expression, and mitogenesis in rat mesangial cells. J Clin Invest 83: 708–712
Sogabe K, Nirei H, Shoibo M, Nomoto A, Ao S, Notsu Y, Ono T (1993) Pharmacological profile of FR 139317, a novel, potent endothelin ET_A receptor antagonist. J Pharmacol Exp Ther 264: 1040–1046
Sorensen SS, Madsen JK, Pedersen EB (1994) Systemic and renal effect of intravenous infusion of endothelin-1 in healthy human volunteers. Am J Physiol 266: F411–F418
Stein PD, Hunt JT, Floyd DM, Moreland S, Dickinson KEJ, Mitchell C, Liu ECK, Webb ML, Murugesan N, Dickey J, McCullen D, Zhang R, Lee VG, Serafino R, Delaney, Schaeffer, TR, Kozlowski M (1994) The discovery of sulfonamide endothelin antagonists and the development of the orally active ET_A antagonist 5-(dimethylamino)-*N*-(3,4-dimethyl-5-isoxazolyl)-1-naphthalenesulfonamide. J Med Chem 37: 329–331

Takahashi K, Katoh T, Kukunag M, Badr KF (1993) Studies on the glomerular microcirculatory actions of manidipine and its modulation of the systemic and renal effects of endothelin. Am Heart J 125: 609–619

Tanaka T, Tsukuda E, Nozawa M, Nonaka H, Ohno T, Kase H, Yamada K, Matsuda Y (1994) RES-701-1: a novel, potent endothelin ET_B receptor-selective antagonist of microbial origin. Mol Pharmacol 45:724–728

Teerlink JR, Loeffler B, Hess P, Maire J, Clozel M, Clozel J (1994) Role of endothelin receptor antagonist Ro 47-0203 (bosentan). Circulation 90:2510–2518

Télémaque S, Gratton JP, Claing A, D'Órléans-Juste P (1993) Endothelin-1 induces vasoconstriction and prostacyclin release via the activation of endothelin ET_A receptors in the perfused rabbit kidney. Eur J Pharmacol 237:275–281

Tomita K, Ujiie K, Nakanishi T, Tomita S, Matsuda O, Ando K, Shichiri M, Hirata Y, Marumo F (1989) Plasma endothelin levels in patients with acute renal failure. N Engl J Med 324:1127

Wagner K, Albrecht S, Neumayer HH (1987) Prevention of posttransplant acute tubular necrosis by the calcium antagonist diltiazem: a prospective randomized study. Am J Nephrol 7:287–291

Warner TD, Battistini B, Allcock GH, Vane JR (1993) Endothelin ET_A and ET_B receptors mediate vasoconstriction and prostanoid release in the isolated kidney of the rat. Eur J Pharmacol 250:447–453

Weitzberg E, Ahlborg G, Lundberg JM (1991) Long-lasting vasoconstriction and efficient regional extraction of endothelin-1 in human splanchnic and renal tissues. Biochem Biophys Res Commun 180:1298–1303

Wellings RP, Warner TD, Thiemermann C, Cristol JP, Corder R, Vane JR (1993) Vasoconstriction in the rat kidney induced by endothelin-1 is blocked by PD 145 065. J Cardiovasc Pharmacol 22 [Suppl 8]:S103–S106

Wu-Wong JR, Chiou WJ, Magnuson SR, Opgenorth TJ (1994) Endothelin receptor agonists and antagonists exhibit different dissociation characteristics. Biochim Biophys Acta 1224:288–294

Yanagisawa M, Kurihara H, Kimura S, Tombe Y, Kobayashi M, Yazaki Y, Goto K, Masaki T (1988) A novel potent vasoconstrictor peptide produced by vascular endothelial cells. Nature 332:411–415

Yukimura T, Yamashita Y, Miura K, Kim S, Iwao H, Takai M, Okada T (1994) Renal vasodilatng and diuretic actions of a selective endothelin ET_B receptor agonist, IRL 1620. Eur J Pharmacol 264: 399–405

Zeidel ML, Brady HR , Kone BC, Gullans SR, Brenner (1989) Endothelin, a peptide inhibitor of Na^+-K^+-ATPase in intact renal tubular epithelial cells. Am J Physiol 257:C1101–C1107

Zimmerman RS, Maymind M, Barbee RW (1994) Endothelin blockade lowers total peripheral resistance in hemorrhagic shock recovery. Hypertension 23:205–210

The Endothelium and the Pathogenesis of Chronic Vascular Diseases: The Protective Role of Calcium Antagonists

H. Haller

Studies on the pathogenesis of essential hypertension and the development of atherosclerosis have for a long time concentrated on the disturbed contractility or an increased proliferation propensity of smooth muscle cells [1, 2]. More recent work has extended the pathophysiological considerations to other cell systems of the vessel wall which have been recognized to be involved in the pathogenesis of chronic vascular diseases [3, 4]. It has become more and more apparent that the endothelium plays a crucial role in the regulation of the vessel wall under physiological and pathological conditions [5]. The endothelium lines all vessels of the body and is the most important structure for communication between the bloodstream and the vessel wall with smooth muscle cells and matrix proteins [6]. It serves as a barrier that prevents noxious agents from entering the vessel wall. Healthy endothelium has antithrombotic properties that inhibit the adhesion of blood cells (thrombocytes, erythrocytes, and leukocytes) to the vessel wall. Hence, endothelial cells are crucial for maintaining laminar blood flow. Besides its role in the interaction of the vessel wall with cells in the bloodstream, the endothelium fulfills important secretory functions. Endothelial cells can release their secretions into the vessel wall as well as into the bloodstream. Of particular importance for the physiological function of the vessel wall is the vasorelaxing function of endothelial cells [5]. The latter secrete endothelium-derived relaxing factor (EDRF), which relaxes smooth muscle cells [7]. Under physiological conditions, EDRF is permanently released and ensures the patency of normal vessels [10]. In addition, endothelial cells can also release vasoconstrictive factors such as endothelin into the vessel wall [8, 9]. Furthermore, studies over recent years have identified growth factors and chemotactic substances which are produced by damaged or overstimulated endothelial cells and which play a crucial part in structural changes of the vessel wall [11, 12].

The most important of these vasorelaxing factors is nitric oxide (NO) or EDRF. NO is synthesized in endothelial cells from the amino acid L-arginine by NO-synthase [13]. The latter enzyme is continuously active and thus ensures the permanent release of NO. Disturbances of endothelial cell function lower the release of NO and thus reduce vasorelaxation [14, 15] (see Lüscher and Noll this volume). In addition to NO, endothelial cells also release prostaglandins with vasodilative action such as prostacyclin. Prostacyclin induces vasorelaxation and inhibits platelet aggregation [16]. It is possible that the endothelium secretes additional vasorelaxing substances that have not yet been identified. Damaged or excessively activated endothelial cells can also secrete vasconstrictive factors [14, 15], of which the recently discovered endothelin is the best-known representative. In addition, to its strong vasoconstrictive properties, endothelin potentiates the constrictive effects of other vasoactive agonists [17]. Thromboxane is another vasoconstrictor that can be released by damaged endothelial cells [18].

Endothelial cells also release also factors which affect differentiation and growth of vascular smooth muscle cells [11]. The dedifferentiation and subsequent proliferation of vascular smooth muscle cells are important early mechanisms in the pathogenesis of chronic vascular disease. Several factors may influence this process. Increased stress of the vessel wall with enhanced stretching of the vascular smooth muscle cells induce early growth response signals in these cells. In addition, chronic exposure of the vascular smooth muscle cells to vasoconstrictors, such as norepinephrine and angiotensin II, results in dedifferentiation and enhanced growth propensity. The growth propensity of the smooth muscle cells can also be enhanced by growth factors released from endothelial cells such as platelet-derived growth factor (PDGF) or epidermal growth factor. Thus, the endothelium is involved in both the physiological regulation of vasoconstriction and the structural transformation of the vessels under pathologic conditions.

How endothelial cell damage contributes under pathophysiological conditions to vascular disease can be illustrated in diabetes mellitus where hyperglycemia and hyperinsulinemia contribute to the cellular processes occurring in the vessel wall. Patients with diabetes mellitus show pronounced changes in endothelial cell function, which were demonstrated by both in vivo studies and in vitro experiments. Enlarged endothelial cells with electron-dense inclusions as signs of the degenerative changes are seen in small vessels of patients with diabetes mellitus [19, 23]. In addition, there is an increased proliferation of endothelial cells, which is assumed to be a repair process to seal endothelial cell lesions in the wall. Endothelial cell damage in patients with diabetes mellitus is also reflected by markedly elevated plasma concentrations of von Willebrand's factor and factor VIII [15]. The endothelial cell changes in patients with diabetes mellitus are also accompanied by functional disturbances [14]. As already described, the endothelium has a decisive role in regulating the contractility of the vessel wall by secreting vasocactive substances. Damage to endothelial cells initially affects the release of vasorelaxing substances, in particular of EDRF. Several study groups have shown that vasorelaxation is disturbed in animal models of diabetes mellitus, probably because of a decrease in the release of vasorelaxing substances from damaged epithelial cells [21, 22]. The already low secretion of EDRF is further reduced by the elevated arterial blood pressure frequently seen in these patients [20]. Production of vasorelaxing prostacyclin by endothelial cells is likewise disturbed in diabetes mellitus [24, 27]. Endothelial cells also secrete substances with vasoconstrictor action such as endothelin. It was shown in vitro [25] that high glucose concentrations stimulate the secretion of endothelin by aortic endothelial cells. In patients with diabetes mellitus increased circulating levels of endothelin have been observed, indicating that the endothelial cells are damaged [27]. Growth factors such as PDGF, which are secreted by stimulated endothelial cells, also have vasoconstrictor action [28, 29]. In diabetics, the ratio of vasorelaxing to vasconstrictive substances in the vessel wall is thus markedly shifted towards the latter [30].

Role of the Endothelium in the Interaction with Cells in the Bloodstream

Endothelin and other factors produced by endothelial cells are released not only into the vessel wall but also into the bloodstream, where their chemotactic action can induce white blood cells to migrate to the endothelial wall [31, 32]. Near the wall, they slow down and adhere to the endothelial cells. This process of cell adhesion to endothelial cells has been known for a long time and its underlying molecular mechanisms have been identified in recent years. Endothelial cells induce adhesion by expressing specific surface adhesion molecules that can interact with ligands on white blood cells and thrombocytes [33]. Many new adhesion molecules have been discovered in recent years and they are subdivided into three groups: the family of selectins (including adhesion molecules such as ELAM and LECCAM), the group of integrins (among them LFA-1), and a supergene family of immunoglobulins (ICAM-1 and VCAM). These adhesion molecules are expressed upon stimulation of endothelial cells by specific factors such as tumor necrosis factor (TNF-α). Their expression is increased on endothelial cells chronically damaged, for example, by cigarette smoking and hypercholesterolemia, risk factors of atherosclerosis [34–36]. VCAM is expressed on the surface of endothelial cells overlying early "fatty streak" lesions [37]. In rabbits fed cholesterol, the expression of VCAM precedes the subendothelial accumulation of monocytes [38], which suggests a role for VCAM in the early stages of vascular damage [39]. The effect of a high cholesterol diet on VCAM expression in vivo suggests that low density lipoproteins (LDL) or LDL modified by oxidation (oLDL) may be responsible for the induction of VCAM [39]. Cultured endothelial cells incubated with LDL [40] or oLDL [41, 42] showed an increased adhesiveness of monocytes; however, specific adhesion molecules were not studied in these experiments. Kume et al. [43] showed that VCAM expression is enhanced by lysophosphatidyl choline, a component of LDL, and we could recently demonstrate a direct effect of LDL on VCAM expression [44].

It has been shown that under pathophysiological conditions, i.e. high glucose concentration and hyperlipidemia, that not only the endothelium contributes to the enhanced binding of blood cells but that leukocytes and platelets show increased surface expression of proteins, such as integrins (CD 11/18 etc.), which serve as binding proteins for the adhesion molecules [45]. Binding of blood cells to the endothelium affects the pathogenesis of atherosclerosis in several ways [3, 46]. Enhanced binding of leukocytes and platelets leads to changes in the laminar blood flow and to the generation of turbulences. This changes enhance platelet aggregability and thrombosis. The binding of leukocytes to the endothelium also affects production and release of reactive oxygen species from these cells [47]. The adhesion to endothelial cells may possibly affect the release of other substances as well. This has been demonstrated for the binding of monocytes to glycosylated proteins under hyperglycemic conditions. Monocytes have surface receptors for advanced glycosylation end products (AGE) and are activated by binding to glycosylated proteins in the vessel wall [49]. Activation leads to an increased release of cytokines and to growth factors, including PDGF, interleukins, TNF-α, and transforming growth factor β [48]. The stimulating effects of AGE proteins on macrophages have been demonstrated by several in vitro studies (for review see [49]).

Endothelial Permeability

Besides active expression and secretion of growth factors and involvement in the augmented migration of macrophages into the vessel wall, there is a third mechanism by which abnormal endothelial cell function contributes to the increase in smooth muscle cell proliferation and matrix formation. It has already been mentioned that the disturbed permeability of the endothelial layer in patients with diabetes mellitus and/or hyperlipidemia leads to an increased influx of substances from the circulation into the vessel wall [2, 3, 50]. These substances also include growth factors, among them insulin, which is a growth factor for smooth muscle cells [51–53]. The majority of studies investigating the effect of insulin on the growth of smooth muscle cells have been performed in cell cultures. But insulin has also been shown to stimulate DNA synthesis in the rat aorta in vivo. Although insulin alone has only a weak mitogenic activity and no effect at all on migration, it markedly enhances the mitogenic and migratory action of other factors such as PDGF [54]. The greater endothelial permeability of the vessels in patients with hyperinsulinemia can further increase the influx of insulin into the vessel wall. It is therefore conceivable that the epidemiological relationship between hyperinsulinemia and the development of atherosclerosis is mediated by the action of insulin as a growth factor [55].

Endothelial cells also have pronounced antithrombotic properties in vivo. They can actively contribute to local enhancement of coagulation. The factors described above, such as EDRF and prostacyclin, act as inhibitors of thrombocyte activation. Endothelial damage can also affect the role of endothelial cells in fibrinolysis [56]. The expression of plasminogen-activating factors by damaged endothelial cells is markedly reduced [14]. Thus, intact endothelial cells are important not only for the integrity of the vessel wall but also for the interaction with cells in the bloodstream. Endothelial cell dysfunction can lead to accelerated intravascular blood coagulation [57].

Finally, endothelial cells are also able to express and secrete matrix proteins and thus contribute to the sclerotic changes of the vessel wall during the development of atherosclerosis [58]. It has been shown that under high glucose concentrations the expression of collagen IV and fibronectin in endothelial cells is increased and the activity of enzymes involved in collagen synthesis enhanced. It is interesting to note that the hyperglycemia-induced increase in the gene expression of matrix molecules persists for several weeks after restoration of normoglycemia. The persistence of hyperglycemia-induced changes in cell function has also been observed in vitro. Thus, there appears to be not only a brief stimulation of endothelial cells in hyperglycemia but also persistent cell function disturbances due to high glucose concentrations.

From the above-mentioned mechanisms it is evident that the endothelium plays a central role in many of the pathophysiological processes which are involved in the early steps of atherosclerosis. It is therefore of importance to investigate the effects of antiatherosclerotic therapy on endothelial cell function and cell–cell interactions. Despite a large number of experimental and clinical studies on the antiatherosclerotic effects of calcium antagonists, there are few data available on the direct effects of this group of agents on endothelial dysfunction.

Calcium Antagonists and the Endothelium

Calcium antagonists may influence endothelial cell function and its interactions with blood cells by different modes of action. However, up to now only little is known about the direct effects of calcium antagonists on endothelial cell function [79]. In contrast to their effects in vascular smooth muscle, calcium antagonists do not influence intracellular calcium regulation in endothelial cells [61]. Intracellular calcium release and transmembranous calcium influx upon stimulation of endothelial cells by agonists are not affected by calcium antagonists [62]. This is due to the fact that endothelial cells apparently do not possess voltage-operated, L-type calcium channels [61]. Therefore, calcium antagonists do not appear to have any adverse effects on the calcium-dependent formation of NO. On the contrary, there are some reports in the literature that describe an increased NO release by calcium antagonists [62]. However, the exact mechanism of how calcium antagonists may interfere with NO production and release has not yet been elucidated. One possibility is that calcium antagonists interfere with the hypoxia-induced alterations of damaged endothelium [80].

Recently it has been shown that calcium antagonists interfere with the interaction between endothelial and vascular smooth muscle cells [63]. As has been described, activated endothelial cells release endothelin which is a strong activator of vascular smooth muscle contraction. Presumably, this effecct of endothelin on vascular smooth muscle contractility is mostly mediated via the L-type calcium channel [64]. This effect of endothelin on the contractility of vascular smooth muscle can be almost abolished by the concomitant administration of calcium antagonists [65]. Kiowski et al. have demonstrated in patients that calcium antagonists are very effective in preventing the endothelin-induced contraction [66]. These findings indicate that calcium antagonists may be effective drugs under conditions of endothelial dysfunction with increased release of endothelin in the vessel wall.

In addition to the described effects on vascular smooth muscle, we have recently shown that calcium antagonists also inhibit the endothelin-induced activation of macrophages. We observed that endothelin induces release of oxygen free radicals in human macrophages. When these macrophages are incubated with nitrendipine the endothelin-induced O_2 release is markedly diminished [67]. A comparable effect of calcium antagonists on leukocytes has also been shown by other investigators [68, 69].

Calcium Antagonists and Cell–Cell Interaction

The increased binding of blood cells to the vessel wall with subsequent cell activation can also be influenced by calcium antagonists. Firstly, calcium antagonists lead to relaxation of the blood vessel and thereby improve the flow conditions. Laminar flow in relaxed blood vessels diminishes cell–cell interaction. Furthermore, calcium antagonists specifically interact with the different cell types, i. e., platelets, leukocytes, and endothelial cells, which play a role in the pathogenesis of atherosclerosis.

Calcium Antagonists and Platelets

Many investigations have shown that calcium antagonists of the dihydropyridine type can improve thrombocyte function and have an antiaggregatory effect on blood platelets [70]. In vitro, calcium antagonists lower the agonist-induced elevation of intracellular free calcium in thrombocytes [70, 71]. This has been shown for thrombin, ADP and other proaggregatory substances. This seems to be the mechanism whereby calcium antagonists reduce the thrombin- and ADP-induced aggregation of platelets. It has also been demonstrated in patients that the administration of calcium antagonists results in a slightly decreased aggregation of platelets in vivo [72]. One of the mechanisms contributing to this effect is a lowered release of thromboxane from thrombocytes during treatment with calcium antagonists [70].

Calcium Antagonists and Cell Adhesion

It has recently been demonstrated that calcium antagonists can lower the expression of the adhesion molecules in cultured endothelial cells. The calcium antagonist nitrendipine reduced the surface expression of the adhesion molecule VCAM [44]. Since endothelial cells have no voltage-sensitive (L-type) calcium channels, these findings suggest that calcium antagonists exert effects on endothelial cells which are not dependent on their inhibitory effect on L-type calcium channels [61]. The finding of a decreased surface expression of adhesion molecules by calcium antagonists is consistent with those of McDonagh and Rauzzino, who recently observed a decrease in leukocyte adhesion in the coronary microcirculation after exposure to a calcium antagonist [73]. The mechanisms underlying this effect remain to be elucidated.

Antiatherosclerotic Effect of Calcium Antagonists

It has been speculated since the early 1970s that calcium antagonists might stop the sclerotic process and calcification of the vessel wall. Recently, new concepts of this antiatherosclerotic effect have been proposed [81]. Fleckenstein and coworkers performed many animal studies that provided the basic data for an antiatherosclerotic action of calcium antagonists [74]. Most of the investigations by Fleckenstein's study group were done in rats. Others examined the antiatherosclerotic effects of calcium antagonists in rabbits on a high-cholesterol diet or in genetic models of hypercholesterolemia. All these studies have shown that calcium antagonists (at high doses) can prevent or reduce the deposition of cholesterol or calcium in the vessel wall (for review see [75, 76]).

The results of these animal studies and the cellular mechanisms of calcium antagonist action in the pathogenesis of atherosclerosis described above served as a basis for designing several clinical trials to investigate the possible antiatherosclerotic effects of calcium antagonists in patients. Such clinical studies face a number of serious problems. Since, as already mentioned, atherosclerotic changes develop over years or even decades, it is of course difficult to evaluate the effects of treatment during a relatively short period of time (maximally 1–3 years). Therapeutic benefits of the administered agents might not become apparent during the study period. Another question that arises relates to the vascular region to be used for evaluating the antiatherosclerotic effect. Atherosclerosis does not affect all vascular regions simultaneously, and it is also possible that calcium antagonists act differently in the various vascular regions in accordance with different pathogenetic mechanisms of atherosclerosis in these regions. The INTACT study investigated patients with mild to moderate coronary sclerosis who were treated with nifedipine for 3 years [77]. It was shown that this calcium antagonist primarily had a significant effect on the development of new plaques. A similar result was reported for the Toronto study performed by Waters et al. (78]. Neither of these studies found a significant change in the resolution of already existing plaques. The clinical findings obtained so far are in good agreement with what is known about the cellular and molecular mechanisms involved in the pathogenesis of atherosclerosis and about the site of action of calcium antagonists at the cellular level. Calcium antagonists primarily affect the cellular interactions of endothelial cells, smooth muscle cells, monocytes, and thrombocytes, which play a central role in the early phases of the development of atherosclerosis, whereas their protective effect on the vascular system appears to be low at later stages with increased matrix formation and central necrosis of atherosclerotic plaques. Thus, a therapeutic conclusion to be drawn from the present studies is to use calcium antagonists as prophylactic agents at an early stage in the development of atherosclerotic vascular changes.

Summary

It has been shown in recent years that calcium antagonists – whose blood-pressure-lowering effects have been known for a long time – also play a crucial role in influencing the pathophysiological mechanisms that underlie chronic cardiovascular diseases. Calcium antagonists affect the different cell systems such as vascular smooth muscle cells, endothelial cells, monocytes and macrophages as well as thrombocytes, which play an important part in the development of chronic vascular changes. Studies on the pathogenesis of essential hypertension and the development of atherosclerosis have for a long time concentrated on the disturbed contractility or an increased proliferation propensity of smooth muscle cells. More recent work has extended the pathophysiological considerations to other cell systems of the vessel wall which have been recognized to be involved in the pathogenesis of chronic vascular diseases. It has become more and more apparent that the endothelium plays a crucial role in the regulation of the vessel wall under physiological and pathological conditions. The endothelium lining all vessels of the body and is the most important structure for communication between the flowing blood, on the one hand, and the vessel wall with smooth muscle cells and matrix proteins, on the other. It serves as a barrier that prevents noxious agents from entering the vessel wall. Healthy endothelium has antithrombotic properties that inhibit the adhesion of blood cells to the vessel wall. Besides its role in the interaction of the vessel wall with cells in the flowing blood, the endothelium fulfills important secretory functions. Of particular importance is the secretion of endothelium-derived relaxing factor (EDRF), which relaxes smooth muscle cells. Furthermore, studies over recent years have identified growth factors and chemotactic substances which are produced by damaged or overstimulated endothelial cells and which play a crucial part in structural changes of the vessel wall. Calcium antagonists influence the above-described interactions and endothelial cell function by different modes of action. It has recently been demonstrated that calcium antagonists can lower the expression of the adhesion molecules ICAM-1 and VCAM-1 in cultured endothelial cells. Several studies in patients have shown that the effect of endothelin on smooth muscle can be markedly reduced by calcium antagonists of the dihydropyridine type. Calcium antagonists can improve thrombocyte function and have an antiaggregational effect on blood platelets. Calcium antagonists furthermore influence white blood cells. Studies have shown that calcium antagonists can lower the release of free radicals from these cells. Initial findings suggest that the adhesion of white blood cells in the vascular system is longer during treatment with calcium antagonists. Thus, a therapeutic conclusion to be drawn from the present studies is to use calcium antagonists as prophylactic agents at an early stage in the development of atherosclerotic vascular changes.

References

1. Folkow B (1982) Physiological aspects of primary hypertension. Physiol Rev 62:347–504
2. Gordon D, Reidy MA, Benditt EP, Schwartz SM (1990) Cell proliferation in human coronary arteries. Proc Natl Acad Sci U S A 87:4600–4604
3. Schwartz CJ, Valente AJ, Sprague EA (1993) A modern view of atherogenesis. Am J Cardiol 71 (6):21–29
4. Ross R (1993) The pathogenesis of atherosclerosis: a perspective for the 1990. Nature 362:801–809
5. Lüscher TF, Vanhoutte PM (1990) The endothelium: modulator of cardiovascular function. CRC Press, Boca Raton
6. Dzau VJ, Gibbons GH (1988) Cell biology of vascular hypertrophy in systemic hypertension. Am J Cardiol 62:30G–35G
7. Vallance P, Collier J, Moncada S (1989) Effects of endothelium-derived nitric oxide on peripheral arteriolar tone in man. Lancet i:997–1000
8. Yanagisawa M, Kurihara H, Kimura S, Mitsiu Y, Kobayashi M, Watanabe TX, Masaki T (1988) A novel potent vasoconstrictor peptide produced by vascular endothelial cells. Nature 388:411–415
9. Panza JA, Quyyumi AA, Brush JE Jr, Epstein SE (1990) Endothelium-dependent vascular relaxation in patients with essential hypertension. N Engl J Med 323:22–27
10. Lüscher TF (1990) Imbalance of endothelium-derived relaxing and contracting factors: a new concept in hypertension? Am J Hypertens 3:317–330
11. DiCorleto PE, Fox PL (1988) Growth factor production by endothelial cells. In: Una R (ed) Endothelial cells, vol 2. CRC Press, Boca Raton, pp 51–62
12. Egleme C, Creesier F, Wood JM (1990) Local formation of angiotensin II in the rat aorta. Effect of endothelium. Br J Pharmacol 100:237–240
13. Palmer RJM, Ashtor DS, Moncada S (1988) Vascular endothelial cells synthesize nitroxide from L-arginine. Nature 333:664–666
14. Lorenzi M, Cagliero E (1991) Pathobiology of endothelial and other vascular cells in diabetes mellitus; call for data. Diabetes 40:653–659
15. Jensen T, Bjerre-Knudsen J, Feldt-Rasmussen B, Deckert (1989) Features of endothelial dysfunction in early diabetic nephropathy. Lancet 1:461–463
16. Guerra R, Brotherton AFA, Goodwin PJ, Clark CR, Armstrong ML, Harrison DG (1989) Mechanisms of abnormal endothelium-dependent vascular relaxation in atherosclerosis: implications for altered autocrine and paracrine functions of DRF. Blood Vessels 26:300–314

17. Yang Z, Richard V, von Segesser L, Stulz P, Turina M, Lüscher TF (1990) Threshold concentrations of endothelin-1 potentiate contractions to norepinephrine and serotonin in human arteries. A new mechanism of vasospasm. Circulation 82:188–195
18. Strano A (1982) thromboxane B_2 formation and platelet sensitivity to prostacyclin in insulin-dependent and independent diabetics. Thromb Res 26:359–370
19. Williams E, Timperley WR, Ward JD, Duckworth T (1980) Electron microscopic studies of vessels in diabetic peripheral neuropathy. J Clin Pathol 33:462–470
20. Epstein M, Sowers JR (1992) Diabetes mellitus and hypertension. Hypertension 19:403–418
21. Meraji S, Jayakody L, Senaratne MP, Thomson AB, Kappagoda T (1987) Endothelium-dependent relaxation in aorta of BB rat. Diabetes 36:978–981
22. Tesfamariam B, Brown ML, Deykin D, Cohn RA (1990) Elevated glucose promotes generation of endothelium-derived vasoconstrictor prostanoids in rabbit aorta. J Clin Invest 85:929–932
23. Vracko R (1982) A comparison of the microvascular lesions in diabetic mellitus with those of normal aging. J Am Geriatr Soc 30:201–205
24. Davi G, Rini GB, Averna M, Novo S, Di Fede G, Pinto A, Notarbartolo A, Strano A (1982) Thromboxane B_2 formation and platelet sensitivity to prostacyclin in insulin-dependent and independent diabetics. Thromb Res 26:359–370
25. Yamouchi T, Ohnaka K, Takayanagi R, Umeda F, Nawata H (1990) Enhanced secretion of endothelin-1 by elevated glucose levels from cultured bovine aortic endothelial cells. FEBS Lett 267:16–18
26. Takahashi K, Ghatei MA, Lam HC, O'Halloran DJ, Bloom SR (1990) Elevated plasma endothelin in patients with diabetes mellitus. Diabetologia 33:306–310
27. Butkus A, Srinska VA, Schumacher OP (1980) Thromboxane production and platelet aggregation in diabetic subjects with clinical complications. Thromb Res 19:211–223
28. Berk BC, Alexander RW, Brock TA, Gimbrone MA, Webb RC (1986) Vasoconstriction – a new role for platelet-derived growth factor. Science 232:87–90
29. Pober JS, Cotran RS (1990) Cytokines and endothelial cell biology. Physiol Rev 70:427–451
30. Panza JA, Quyyumi AA, Brush JE Jr, Epstein SE (1990) Endothelium-dependent vascular relaxation in patients with essential hypertension. N Engl J Med 323:22–27
31. Simionescu M, Simionescu N (1986) Functions of the endothelial cell surface. Annu Rev Physiol 48:279–304
32. Scharf RE, Harker LA (1987) Thrombosis and atherosclerosis: regulatory role of interactions among blood components and endothelium. Blut 55:131
33. Springer TA (1990) The sensation and regulation of interactions with the extracellular environment: the cell biology of lymphocyte adhesion receptors. Annu Rev Cell Biol 6:359–402
34. DiCorleto PE, Chisolm GM (1986) Participation of the endothelium in the development of the atherosclerotic plaque. Prog Lipid Res 25:3655–3674
35. Faggiotto A, Ross R, Harker L (1986) Studies of hypercholesterolemia in the nonhuman primate. I. Changes that lead to fatty streak formation. Arteriosclerosis 4:323–340
36. Gerrity RG (1981) The role of monocyte in atherogenesis. 1. Transition of bloodborne monocytes into foam cells in fatty lesions. Am J Pathol 103:181–190
37. Cybulsky MI, Gimbrone MJ (1991) Endothelial expression of a mononuclear leukocyte adhesion molecule during atherogenesis. Science 251 (4995):788–791
38. Li H, Cybulsky MI, Gimbrone MJ, Libby P (1993) An atherogenic diet rapidly induces VCAM-1, a cytokine-regulatable mononuclear leukocyte adhesion molecule, in rabbit aortic endothelium. Arterioscler Thromb 13 (2):197–204
39. Libby P, Hansson GK (1991) Involvement of the immune system in human atherogenesis: current knowledge and unanswered questions. Lab Invest 64:5–15
40. Alderson LM, Endemann G, Lindsey S, Pronczuk A, Hoover RI, Hayes KC (1986) LDL enhances monocyte adhesion to endothelial cells in vitro. Am J Pathol 123:334–341
41. Berliner JA, Territo MC, Sevanian A et al (1990) Minimally modified LDL stimulates monocyte endothelial interactions. J Clin Invest 85:1260–1267
42. Frostegard J, Haegerstrand A, Gidlund M, Nilsson J (1991) Biologically modified LDL increases the adhesive properties of endothelial cells. Atherosclerosis 90:119–126
43. Kume N, Cybulsky MI, Gimbrone MJ (1992) Lysophosphatidylcholine, a component of atherogenic lipoproteins, induces mononuclear leukocyte adhesion molecules in cultured human and rabbit arterial endothelial cells. J Clin Invest 90 (3):1138–1144

44. Haller H, Schaper D, Philipp S, Distler A (1995) LDL induces surface expression of adhesion molecules ICAM-1 and VCAM on endothelial cells via the activation of protein kinase C. Hypertension 25:511–517
45. Gilcrease MZ, Hoover RL (1991) Examination of monocyte adherence to endothelium under hyperglycemic conditions. Am J Pathol 139:1089–1097
46. DiCorleto PE (1993) Cellular mechanisms of atherogenesis. Am J Hypertens 6:314S–318S
47. Schaberg T, Rau M, Kaiser D, Fassbender M, Lode H, Haller H (1992) Increased number of alveolar macrophages expressing adhesion molecules of the leukocyte adhesion molecule family in smoking subjects: association with cell-binding ability and superoxide anion production. Am Rev Respir Dis 146:1287–1293
48. Vlassara H, Brownlee M, Monogue K, Dinarello CA, Pasagian A (1988) Cachectin/TNF and IL-1 induced by glucose-modified proteins: role in normal tissue remodeling. Science 240:1546–1548
49. Brownlee M, Cerami A, Vlassara H (1988) Advanced glycosylation end products in tissue and the biochemical basis of diabetic complications. N Engl J Med 318:1315–1321
50. Esposito C, Gerlach H, Brett J, Stern D, Vlassara H (1989) Endothelial receptor-mediated binding of glucose-modified albumin is associated with increased monoplayer permeability and modulation of cell surface coagulant properties. J Exp Med 170:1387–1407
51. Stout RW, Bierman EL, Ross R (1975) Effect of insulin on the proliferation of cultured primate arterial smooth muscle cells. Circ Res 36:319–327
52. King GL, Goodman AD, Buzney S, Moses A, Kahn CR (1985) Receptors and growth-promoting effects of insulin and insulin-like growth factors on cells from bovine retinal capillaries and aorta. J Clin Invest 75:1028–1036
53. Weinstein R, Stemerman MB, Maciag T (1981) Hormonal requirements for growth of arterial smooth muscle cells in vitro: an endocrine approach to atherosclerosis. Science 212:818–820
54. Capron L, Jarnet J, Kazandjian S, Housset E (1986) Growth-promoting effects of diabetes and insulin on arteries: an in vivo study of rat aorta. Diabetes 35:973–978
55. Stout RW (1990) Insulin and atheroma. 20-yr perspective. Diabetes Care 13:631–654
56. Shultz PJ, Raij L (1990) Endogenously synthesized nitric oxide prevents endotoxin-induced glomerular thrombosis. J Clin Invest 90:1718–1725
57. Steiner G (1981) Diabetes and atherosclerosis: an overview. Diabetes 1981; 30 [Suppl 2]:1–7
58. Cagliero E, Maiello M, Boeri D, Roy S, Lorenzi M (1988) Increased expression of basement membrane components in human endothelial cells cultured in high glucose. J Clin Invest 82: 735–738
59. Cagliero E, Roth T, Roy S, Lorenzi M (1991) Characteristics and mechanisms of high-glucose-induced over expression of basement membrane components in cultured human endothelial cells. Diabetes 40:102–110
60. Roy S, Sala R, Cagliero E, Lorenzi M (1990) Over expression of fibronectin induced by diabetes or high glucose: phenomenon with a memory. Proc Natl Acad Sci U S A 87:404–408
61. Himmel HM, Whorton AR, Strauss HC (1993) Intracellular calcium, currents, and stimulus – response coupling in endothelial cells. Hypertension 21:112–127
62. Gunther J, Dhein S, Rosen R, Klaus W, Fricke U (1992) Nitric oxide (EDRF) enhances the vasorelaxing effect of nitrendipine in various isolated arteries. Basic Res Cardiol 87:452–460
63. Lüscher TF, Espinosa E, Dubey RK, Yang Z (1993) Vascular biology of human coronary artery and bypass graft disease. Curr Opin Cardiol 8:963–974
64. Yang Z, Bauer E, von Segesser L, Stulz P, Turina M, Lüscher TF (1990) Different mobilization of calcium in endothelin-i – induced contractions in human arteries and veins: effects of calcium antagonists. J Cardiovasc Pharmacol 16:654–660
65. Goto K, Kasuya Y, Matsuki N, Takuwa Y, Kurihara H, Kimura S, Yanagisawa M, Masaki T (1989) Endothelin activates the dihydropyridine – sensitive, voltage-dependent calcium channel in vascular smooth muscle. Proc Natl Acad Sci U S A 86:3915–3918
66. Kiowski W, Lüscher TF, Linder L, Bühler FR (1991) Endothelin-1 induced vasoconstriction in Man: reveral by calcium channel blockade but not by nitrovasodilators or EDRF. Circulation 83:469–475
67. Haller H, Schaberg T, Lindschau C, Quass P, Lode H, Distler A (1991) Endothelin increases intracellular free calcium, protein phosphorylation and O_2-production in human alveolar macrophages. Am J Physiol 261:L723–L713

68. Wright B, Zeitman I, Greig R, Poste G (1985) Inhibition of macrophage function by calcium channel blockers and calmodulin antagonists. Cell Immunol 95:46–53
69. Jouvin-Marche E, Cerrina J, Coeffier E, Duroux P, Benviste J (1983) Effect of the calcium antagonist nifedipine on the relase of PAF, slow-reacting substance and beta-glucuronidase from human neutrophils. Eur J Pharmacol 89:19–26
70. Tschöpe D, Kaufmann L, Roesen P, Ohlrogge R, Gries FA (1988) Influence of a single dose of nitrendipine on whole platelet activity in healthy subjects. J Cardiovasc Pharmacol 12:S167–S269
71. Rostagno C, Abbate R, Gensini GF, Coppo M, Prisco D, Boddi M, Serneri GG (1991) In vitro effects of two novel calcium antagonists (nitrendipine and nisoldipine) on intraplatelet calcium distribution, platelet aggregation and thromboxane A2 formation. Thromb Res 63:457–462
72. Haller H, Lenz T, Lüdersdorf M, Distler A, Philipp T (1987) Changes in sensitivity to angiotensin II in platelets. J Cardiovasc Pharmacol 10 [Suppl 10]:S44–S46
73. McDonagh PF, Rauzzino MJ (1993) Stimulated leukocyte adhesion in coronary microcirculation is reduced by a calcium antagonist. Am J Physiol 265:H476–H483
74. Fleckenstein-Grun G, Frey M, Thimm F, Hofgartner W, Fleckenstein A (1992) Calcium overload – an important cellular mechanism in hypertension and arteriosclerosis. Drugs 1:23–30
75. Haller H (1993) Calcium antagonists and cellular mechanisms of glomerulosclerosis and atherosclerosis. Am J Kidney Dis 21:26–31
76. Borcherding SM, Meeves SG, Klutman NE, Howard PA (1993) Calcium-channel antagonists for prevention of atherosclerosis. Ann Pharmacother 27:61–67
77. Lichtlen PR, Hugenholtz PG, Rafflenbeul W (1990) Retardation of angiographic progression of coronary artery disease by nifedipine. Results of the International Nifedipine Trial of Antiatherosclerotic Therapy (INTACT). Lancet 335:1109–1113
78. Waters D, Lesperance J, Francetich M (1990) A controlled clinical trial to assess the effect of a calcium channel blocker on the progression of coronary atherosclerosis. Circulation 82:1940–1953
79. Ciriaco E, Abbate F, Ferrante F, Laura R, Amenta F (1993) Structural changes in the endothelium of the femoral artery of spontaneously hypertensive rats: sensitivity to isradipine treatment. J Hypertens 11:515–522
80. Wilkie ME, Stevens CR, Cunningham J, Blake D (1992) Hypoxia-induced von Willebrand factor release is blocked by verapamil. Miner Electrolyte Metab 18:141–144
81. Collins P, Rosano GM, Jiang C, Lindsay D, Sarrel PM, Poole-Wilson PA (1993) Cardiovascular protection by oestrogen – a calcium antagonist effect? Lancet 341:1264–1265

Insulin and the Regulation of the Cardiovascular System: Role of the L-arginine Nitric Oxide Pathway and the Sympathetic Nervous System

U. Scherrer

Introduction

The past several years have witnessed a major surge of interest in the cardiovascular actions of insulin and the potential role played by both the sympathetic nervous system and the L-arginine nitric oxide system in the mediation of these actions.

This interest was stimulated on the one hand, by epidemiological studies that demonstrated an association between obesity, insulin resistance, and hypertension (Modan et al. 1985; Lucas et al. 1985). Subsequently, it has been suggested that many lean subjects with essential hypertension also have insulin resistance (Ferrannini et al. 1987; Bonora et al. 1987). Insulin resistance does not seem to be secondary to increased blood pressure per se because it persists during antihypertensive treatment, and it is not found in many secondary forms of hypertension (Shen et al. 1988). Moreover, insulin resistance is restricted to the actions of insulin on carbohydrate metabolism, whereas its renal and sympathetic actions are preserved. This latter observation has led the basis for the insulin hypothesis of hypertension (Landsberg 1986; Reaven 1988).

On the other hand, and more recently, insulin's vascular actions have become the focus of much interest because evidence has accumulated, suggesting that they may play a role in the promotion of insulin delivery to its main target tissue, namely skeletal muscle. Indeed, to gain access to its target cell, insulin must travel from the portal vein to skeletal muscle capillaries, cross the capillary barrier, and then diffuse through the interstitial space. Two types of observations have led the basis for the concept of a vascular origin of insulin resistance (Baron 1993). First, in Pima Indians, an ethnic group with an extremely high prevalence of insulin-resistance and diabetes, capillary density in skeletal muscle was found to be directly correlated with the rate of insulin-induced stimulation of muscle glucose uptake, suggesting that the diffusion distance between the capillary and the insulin sensitive target cell may represent a rate limiting factor for insulin action (Lillioja et al. 1987). Consistent with this observation, findings in dogs employing measurements of insulin concentration in lymph as a marker of interstitial fluid demonstrated markedly lower insulin concentration in lymph than in plasma, suggesting an action gradient that increases proportionally to the distance from the capillary (Yang et al. 1989). Second, and more recently, evidence has been provided indicating that insulin-induced stimulation of muscle blood flow, by boosting its own delivery to target tissues appears to play an integral part in its main action the promotion of glucose disposal in skeletal muscle tissue (Laakso et al. 1990a). Using the euglycemic clamp technique (DeFronzo et al. 1979)

Laakso and colleagues found that insulin-induced stimulation of muscle glucose uptake was closely related to stimulation of muscle blood flow.

This review will be divided in two parts, one relating to the role played by the sympathetic nervous system and the L-arginine nitric oxide system in mediating insulin's vascular actions in healthy subjects, and the other related to alterations in insulin's vascular and sympathetic actions in insulin-resistant states such as obesity.

Vascular and Sympathetic Actions of Insulin in Lean Subjects

Historical Background

Insulin, apart from its effects on intermediary metabolism, also has effects on the heart and the vasculature that are thought to be mediated at least in part by the sympathetic nervous system. Studies in experimental animal models have provided evidence consistent with the hypothesis that acute insulin administration may stimulate sympathetic outflow by a central mechanism (Pereda et al. 1962; Sauter et al. 1983). In humans, insulin infusion has been shown to stimulate sympathetic activity as determined both by measurements of plasma catecholamines (Rowe et al. 1981; Anderson et al. 1991; Berne et al. 1992; Lembo et al. 1992; Scherrer et al. 1993; Vollenweider et al. 1993), and, more recently, by direct microneurographic recordings of sympathetic nerve action potentials targeted at the skeletal muscle vasculature (Anderson et al. 1991; Berne et al. 1992; Scherrer et al. 1993; Vollenweider et al. 1993). This insulin-induced sympathetic activation in skeletal muscle is thought to be, at least in part sympathetic vasoconstrictor, because it has been shown to be accompanied by muscle norepinephrine release (Lembo et al. 1992). Surprisingly, however, there is abundant evidence that insulin stimulates muscle blood flow and decreases, not increases, vascular resistance in skeletal muscle. Indeed, shortly after the introduction of insulin into clinical practice, hyperinsulinemia-induced vasodilation was first reported (Abramson et al. 1939). In these early studies using injection of massive doses of insulin, the vasodilatory effect was attributed, at least in part, to the epinephrine release evoked by insulin-induced hypoglycemia (DiSalvo et al. 1956). However, hyperinsulinemia-induced stimulation of muscle blood flow is not dependent on epinephrine release, because it has been shown to occur both in adrenalectomized humans (Ginsburg and Paton 1956) and during euglycemic hyperinsulinemia, situations in which epinephrine does not increase. Since then, numerous studies in both experimental animal models (Liang et al. 1982) and humans (Bennet et al. 1990; Laakso et al. 1990a; Anderson et al. 1991; Scherrer et al. 1993; Vollenweider et al. 1993) have confirmed this insulin-induced stimulation of muscle blood flow.

In the following, we will review recent data that have provided important new insight in the underlying mechanisms by which insulin exerts its vascular and sympathetic effects in lean healthy humans.

Differential Effects of Hyperinsulinemia and Carbohydrate Metabolism on Sympathetic Nerve Activity and Muscle Blood Flow

At the beginning of the present decade, independent findings by Baron and colleagues (Laakso et al. 1990a, b, 1992a, b), Anderson and colleagues (Anderson et al. 1991), and Swedish investigators (Berne et al. 1992) firmly established the concept that in lean subjects, euglycemic hyperinsulinemia at high physiological concentrations evokes marked increases in both sympathetic nerve activity and blood flow in skeletal muscle tissue. The underlying mechanism however remained unknown. Specifically, it was not clear whether the primary stimulus which triggers sympathetic activation and vasodilation in human skeletal muscle is hyperinsulinemia per se, or hyperinsulinemia-induced stimulation of carbohydrate metabolism.

To address this issue, Vollenweider and colleagues (1993) performed simultaneous microelectrode recordings of sympathetic nerve discharge to calf muscles, plethysmographic measurements of calf blood flow in the contralateral limb, together

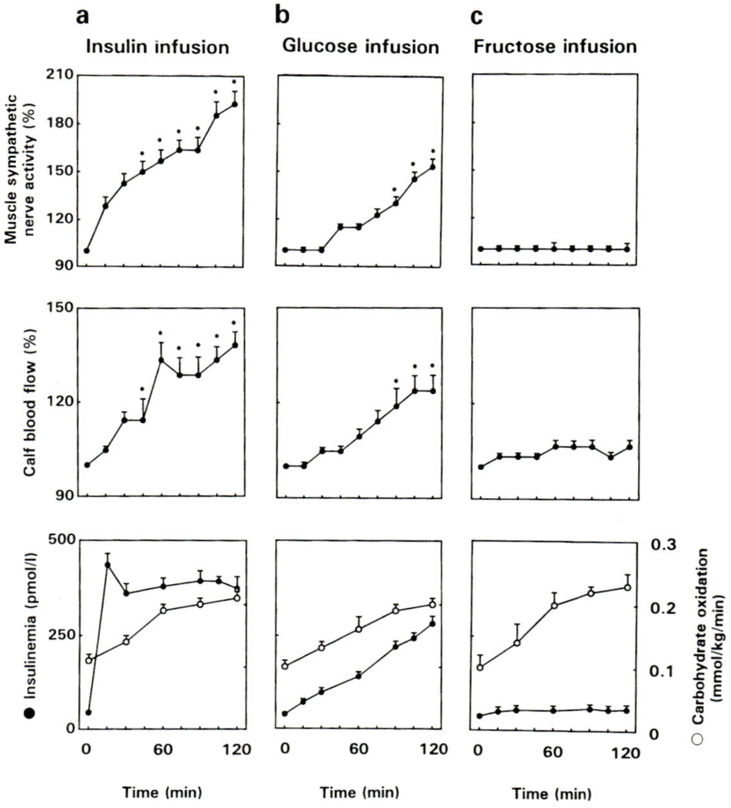

Fig. 1a–c. Effects on plasma insulin concentration, carbohydrate oxidation rate, calf blood flow, and muscle sympathetic nerve activity of 2-h infusions of: **a** insulin/glucose (euglycemic hyperinsulinemic clamp), **b** glucose or **c** fructose. Data are means ± SE for six subjects (fructose $n = 5$). * $p < 0.05$ vs. corresponding baseline value. (From Vollenweider et al. 1993)

with calorimetric determinations of carbohydrate oxidation during experimental interventions in which insulinemia was altered independently of carbohydrate metabolism (euglycemic hyperinsulinemic clamp, and exogenous glucose or fructose infusion). They found that for comparable rates of carbohydrate metabolism and oxidation, insulin/glucose infusion that resulted in a twofold larger increase in plasma insulin concentration than did glucose infusion alone, also evoked twofold larger increases in both muscle sympathetic nerve activity and calf blood flow (Fig. 1). Furthermore, fructose infusion that increased carbohydrate oxidation to a rate similar to that observed during insulin/glucose infusion, but had only a barely detectable effect on plasma insulin concentration, did not stimulate sympathetic activity and calf blood flow. Thus, this elegant series of experiments allowed to establish the concept that sympathetic activation and vasodilation in skeletal muscle are related to hyperinsulinemia per se rather than to stimulation of carbohydrate metabolism.

In addition, and consistent with other reports (Anderson et al. 1991; Berne et al. 1992), these studies added further evidence to the observation that in lean healthy subjects, acute hyperinsulinemia at high physiological concentrations does not raise arterial blood pressure. This finding may be related to a balance between insulin's opposing sympathoexcitatory pressor and vasodilatory depressor actions. Finally, regarding the underlying mechanism(s) by which insulin exerts its sympathoexcitatory and vasodilatory effects, the observation during insulin/glucose infusion (Fig. 1) that sympathetic activity and calf blood flow did not show a peak increase during the initial peak increase in of plasma insulin concentration, and showed a further increase when plasma insulin levels were at steady state, suggest that insulin may have to reach the interstitial space to exert its excitatory effects. One potential site of action is the central nervous system (Pereda et al. 1962; Margolis and Altszuler 1967; Van Houten and Posner 1979; Sauter et al. 1983; Lembo et al. 1992).

Effects of Dexamethasone Administration on Insulin-Induced Sympathetic Activation and Vasodilation

In experimental animals there is compelling evidence that the modulatory effects of insulin on food intake are centrally mediated (Schwartz et al. 1992). Furthermore, it has been shown in rats that glucocorticoids impair the ability of insulin to regulate food intake by altering insulin-induced central neural peptide and neurotransmitter release (Gosink et al. 1992). Interestingly, some of these peptides such as corticotropin releasing hormone (CRH) and neuropeptide Y (NPY) also have cardiovascular effects (Kurosawa et al. 1985; Scott et al. 1989; Overton and Fisher 1989). For example, in rats, central administration of CRH evokes sympathetic nerve (Brown and Fisher 1985; Kurosawa et al. 1985) and hind-limb blood flow (Overton and Fisher 1989) responses that closely mimic the insulin-induced sympathoexcitatory and vasodilator responses in humans. Conversely, dexamethasone administration suppresses stress-induced sympathetic activation in experimental animals (Brown and Fisher 1985), presumably by acting on release of CRH (Hohnloser et al. 1989) or related central autonomic nervous system regulating peptides (Gruber and Callahan 1989). These observations raise the possibility that dexamethasone administration may attenuate the sympathetic and vascular effects of insulin.

To test this concept, we have performed simultaneous microelectrode recordings of sympathetic nerve discharge to skeletal muscle and plethysmographic measurements of calf blood flow during insulin/glucose infusion in lean healthy subjects (Scherrer et al. 1993). Each subject received two infusions, one after short-term administration of dexamethasone (2 mg/day) and one after administration of a placebo for 48 h. We found that plasma insulin levels within the high physiological range that consistently elicited large increases in sympathetic outflow and blood flow in skeletal muscle had no such stimulatory effect after dexamethasone administration (Fig. 2). These findings provide evidence that dexamethasone dramatically impairs the ability of insulin to stimulate sympathetic activity and blood flow in humans. This observation is not related to a nonspecific impairment of sympathetic and vasodilator responsiveness by dexamethasone, because reflex sympathetic responses to the Val-

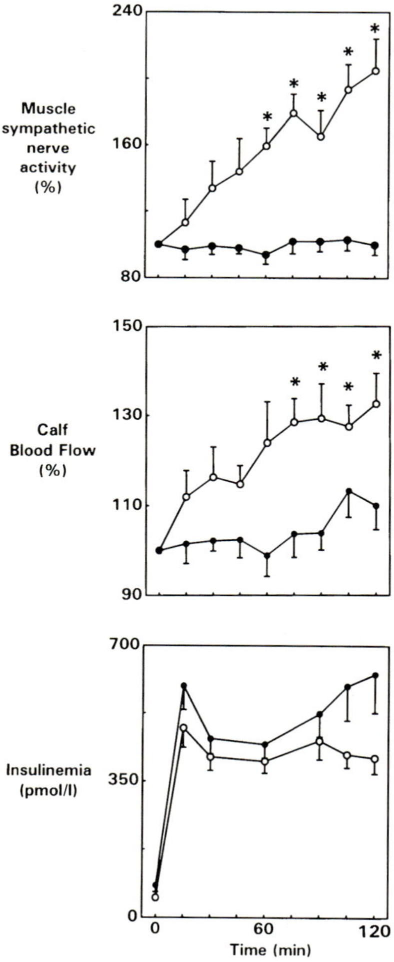

Fig. 2. Comparative effects of insulin infusion (6 pmol/kg per minute) performed alone (*open circles*) and performed after dexamethasone administration (2 mg/day) for 48 h (*closed circles*) in six lean subjects. Data are means ± SE for six subjects. * $p < 0.05$ vs insulin infusion alone. (From Scherrer et al. 1993)

salva maneuver (Scherrer et al. 1990) and immersion of the hand in ice water (Victor et al. 1987) were preserved after dexamethasone, indicating that efferent sympathetic pathways were not altered by this agent and could respond appropriately to baroreceptor deactivation and stimulation of cutaneous afferents. Finally, dexamethasone also did not alter vascular responsiveness to reactive hyperemia, indicating preservation of blood vessel responsiveness to this noninsulin vasodilator stimulus (Takeshita and Mark 1980). Taken together our findings suggest that insulin-induced stimulatory effects on sympathetic activity and blood flow in skeletal muscle could be centrally mediated and could involve the release of specific neuropeptides. CRH is such a candidate peptide.

In this regard, Randin and colleagues (1995) have recently provided additional evidence suggesting that CRH could play an important role in the regulation of sympathetic outflow in humans. They examined sympathetic and cardiovascular effects of alcohol infusion in normal subjects. Each subject received two infusions, one after short term administration of dexamethasone and one after administration of placebo. Alcohol infusion alone evoked marked and progressive increases in sympathetic nerve discharge that were accompanied by an increase in arterial pressure. After dexamethasone, alcohol infusion had no detectable sympathetic effect and was now associated with a decrease in arterial pressure. Thus, these finding indicate that alcohol induces pressor effects by sympathetic activation that appear to be centrally mediated and could involve CRH release.

So far we have focused on effects of high physiological insulin concentrations on sympathetic outflow and muscle blood flow and provided evidence that these effects are related directly to insulin per se (rather than to stimulation of carbohydrate metabolism) and appear to be centrally mediated (possibly involving CRH release). An important unresolved issue with regard to insulin resistant states (and potential pathophysiological consequences of insulin-induced sympathetic activation in such states) was however, whether low dose insulin infusion (resulting in plasma insulin concentrations commonly observed in fasting insulin-resistant subjects) also has sympathoexcitatory (and vasodilatory) effects.

Effects of Low Dose Insulin Infusion on Sympathetic Activity and Muscle Blood Flow

Vollenweider and colleagues (1994) have recently compared sympathetic and vasodilatory responses in skeletal muscle to three different rates of insulin glucose infusion. On one occasion insulin was infused at a rate of 1 mU/kg per minute (high dose) for 2 h, on another occasion at a rate of 0.5 mU/kg per minute (intermediate dose), and on a third occasion at a rate of 0.15 mU/kg per minute. The major finding of these studies was that insulin infusion at a rate six times and two times lower than the one used during high dose infusion markedly stimulated sympathetic activity and blood flow in skeletal muscle (Fig. 3). Indeed, increases in sympathetic activity were not statistically different during the three experimental conditions.

These findings indicate that even very small increases in plasma insulin concentrations have important effects on sympathetic nerve activity. This observation could be of importance in pathophysiological states characterized by sustained increases in

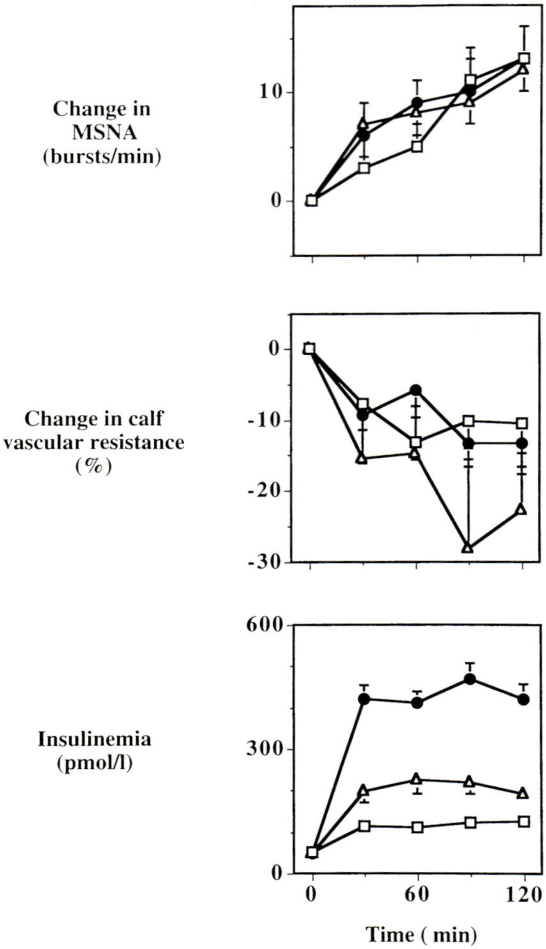

Fig. 3. Effects on plasma insulin concentration, and changes in calf vascular resistance and muscle sympathetic nerve activity (MSNA) of 2-h insulin infusions in lean subjects performed at three different rates (●—●, 1 mU/kg per minute; △—△, 5 mU/kg per minute; □—□, 0.15 mU/kg per minute). Data are means ± SE for five subjects. (From Vollenweider et al. 1994)

plasma insulin concentrations such as obesity (see "Vascular and Sympathetic Actions of Insulin in Insulin-Resistant States" below). In the following we will focus on potential mechanisms underlying insulin-induced vasodilation in skeletal muscle.

Effects of Adrenergic and Cholinergic Blockade on Insulin-Induced Stimulation of Muscle Blood Flow

We have shown so far that insulin is a potent stimulus to both sympathetic outflow and blood flow in skeletal muscle. The consequences of such neural activation on the regulation of muscle blood flow, however, until recently were poorly understood. To examine whether activation of neural vasodilator outflow contributes to, or activation of neural vasoconstrictor outflow limits insulin-induced stimulation of muscle

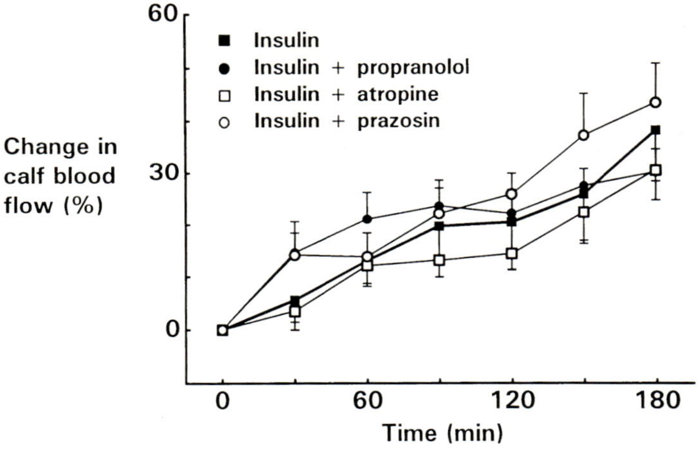

Fig. 4. Effects on calf blood flow of a 3-h insulin/glucose infusion performed alone and performed during concomitant β-adrenergic, cholinergic, or alpha-adrenergic blockade. Data are means ± SE for eight subjects. (From Randin et al. 1994)

blood flow, we have recently examined effects of pharmacological vascular receptor blockade on muscle blood flow during euglycemic hyperinsulinemia (Randin et al. 1994). The major new finding of this study was that in lean subjects, a 3 h insulin/glucose infusion which increased plasma insulin concentrations to high physiological levels evoked marked increases in calf blood flow (and decreases in calf vascular resistance) that were not significantly altered by concomitant administration of either adrenergic or cholinergic antagonists (Fig. 4). These findings indicate that insulin-induced stimulation of muscle blood flow does not result primarily from stimulation of cholinergic or ß-adrenergic mechanisms. Furthermore, they demonstrate that in lean subjects, hyperinsulinemia-induced stimulation of sympathetic outflow in skeletal muscle, which is thought to be at least in part vasoconstrictor, because it is accompanied by increases in plasma norepinephrine concentrations (Anderson et al. 1991; Vollenweider et al. 1993; Scherrer et al. 1993) and local norepinephrine release (Lembo et al. 1992), does not appear to limit insulin-induced vasodilation in the calf vasculature. These findings are consistent with previous observations in conscious dogs indicating that in the absence of hypoglycemia, propranolol infusion does not limit, and phentolamine infusion does not potentiate insulin-induced stimulation of hindlimb blood flow (Liang et al. 1982).

Effects of Inhibition of Nitric Oxide Release on Insulin-Induced Stimulation of Muscle Blood Flow

Over the past few years, endothelium-derived relaxing factor, whose activity is accounted for by nitric oxide (NO), has emerged as a major regulatory mechanism of peripheral vascular tone (Vallance et al. 1989). This factor which is synthesized from the amino acid L-arginine by the enzyme NO-synthase is a potent vasodilator. N^G-monomethyl-L-arginine (L-NMMA), an analogue of L-arginine, by its action as a stereospecific inhibitor of NO-synthase, inhibits both basal and stimulated NO synthesis. For example, in the human forearm, intraarterial L-NMMA infusion causes vasoconstriction, indicating that basal NO-release is an important determinant of resting blood flow (Vallance et al. 1989). Furthermore, the forearm vasodilation evoked by local infusion of acetylcholine (Vallance et al. 1989) and vasopressin (Tagawa et al. 1993) is attenuated after infusion of L-NMMA, indicating that stimulation of NO synthesis contributes to the vasodilation evoked by these pharmacological agents. Finally, there is evidence both in experimental animals (Davisson et al. 1993), and humans (Saenz de Tejada et al. 1989) that NO is a mediator of neural nonadrenergic, noncholinergic vasodilation.

We have recently examined the effects on forearm blood flow of local intraarterial L-NMMA infusion performed before and at the end of a 2-h insulin glucose infusion and compared these effects with those of norepinephrine infusion (an NO-independent vasoconstrictor) (Scherrer et al. 1994b). To assess effects of altered insulin-induced vasodilation by L-NMMA on glucose uptake, we also performed calorimetric determinations of carbohydrate metabolism. The major new findings were that the

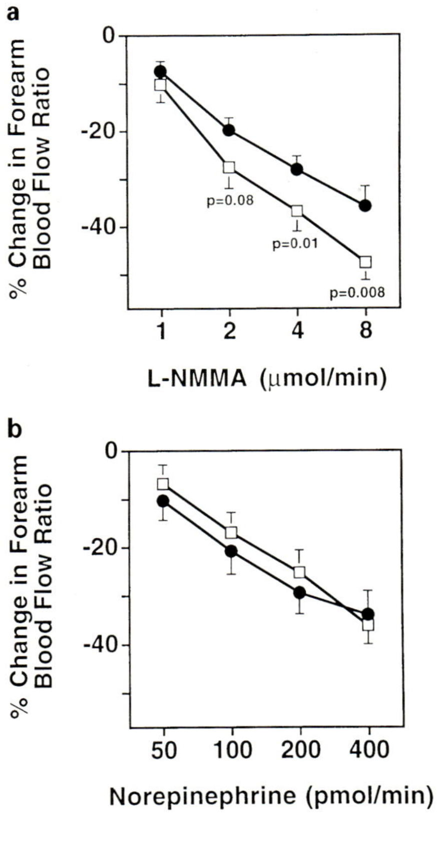

Fig. 5a–b. Percent changes in forearm blood flow ratio (infused/control arm) in response to increasing doses of **a** intraarterial L-NMMA and **b** norepinephrine infusion performed during fasting (*filled circles*) and during euglycemic hyperinsulinemia (*open squares*). Data are means ± SE in six subjects. (From Scherrer et al. 1994b)

decrease in forearm blood flow evoked by L-NMMA administered during insulin/glucose infusion was significantly larger than the one evoked in the absence of insulin infusion, indicating an augmented contribution of NO to overall forearm vascular tone (Calver et al. 1992) during hyperinsulinemia as compared to euinsulinemia (Fig. 5). In contrast, responses to norepinephrine, an NO-independent vasoconstrictor, were similar at basal and during insulin glucose infusion. Second, in all subjects, forearm vasodilation evoked by insulin/glucose infusion was prevented when insulin infusion was preceded by short-term (16 min) intraarterial L-NMMA infusion (Fig. 6). These findings, which are consistent with those of a recent report by Steinberg et al. (1994), demonstrate that insulin's vascular effects in skeletal muscle are mediated by stimulation of nitric oxide release.

We made two additional important observations during these studies. First, as already discussed earlier in this chapter, acute hyperinsulinemia at high physiological concentrations does not increase blood pressure in humans. This observation has been attributed to a balance between the opposing pressor (sympathetic activation) and depressor (vasodilation) effects of insulin in the skeletal muscle bed (Anderson et al. 1991; Vollenweider et al. 1993). Our study provides direct experimental support for this hypothesis and shows that when insulin-induced vasodilation in skeletal muscle is prevented by L-NMMA, blood pressure increases during insulin infusion at high physiological concentrations (Fig. 7).

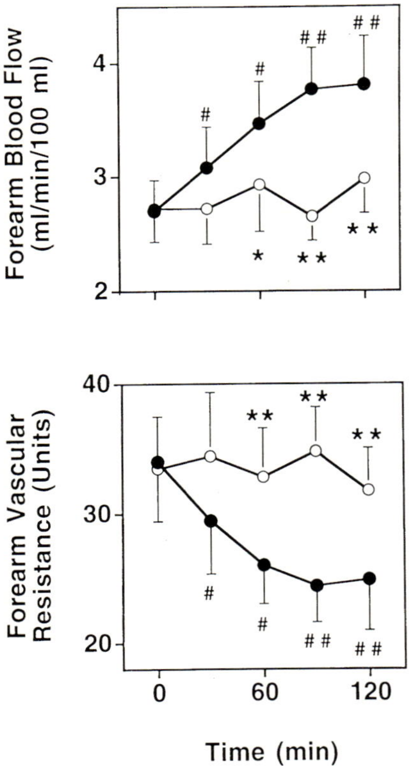

Fig. 6. Forearm vascular responses to 2-h insulin infusion performed alone (*filled circles*) and performed after short term intraarterial L-NMMA (1, 2, 4, and 8 μmol/min for 4 min each) infusion (*open circles*). Data are means ±SE in nine subjects. # $p<0.05$; ## $p<0.005$ vs corresponding time 0; * $p<0.05$; ** $p<0.01$ vs insulin infusion preceded by L-NMMA. (From Scherrer et al. 1994b)

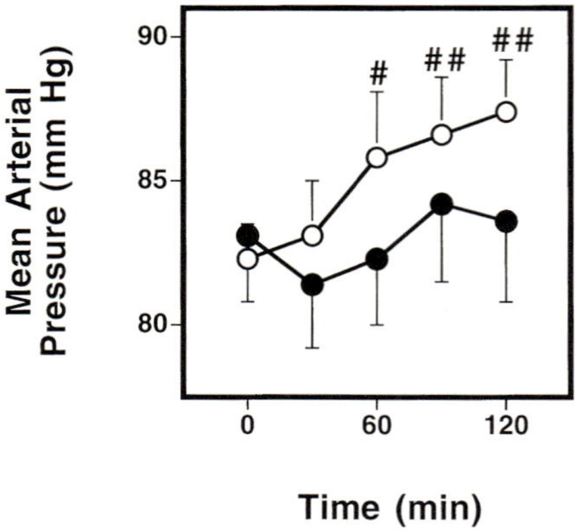

Fig. 7. Blood pressure responses to 2-h insulin infusion performed alone (*filled circles*) and performed after short term intraarterial L-NMMA (1, 2, 4, and 8 μmol/min for 4 min each) infusion (*open circles*). Data are means ±SE for nine subjects. * $p < 0.05$ vs corresponding time 0. (From Scherrer et al. 1994b)

Second, recent findings suggest that there is a direct relationship between insulin-induced stimulation of muscle blood flow and muscle glucose uptake (Laakso et al. 1990a). This observation has led the basis for the vascular hypothesis of insulin resistance (Baron 1993). In our studies, however, suppression of insulin-induced vasodilation did not have any detectable effect on insulin-induced stimulation of whole body glucose uptake or glucose oxidation. This finding challenges the vascular hypothesis of insulin resistance and suggests that insulin-induced stimulation of blood flow per se may not be a primary determinant of muscle glucose uptake and oxidation.

While these data demonstrate a dramatic impairment of insulin-induced vasodilation after L-NMMA administration, they do not allow to determine the underlying mechanism by which insulin stimulates NO release. Recent observations in insulin-resistant humans could provide some clues. There is now convincing evidence that insulin resistance is associated with an impairment in both insulin induced vasodilation (Laakso et al. 1990a; Baron et al. 1993) and sympathetic activation (Vollenweider et al. 1994) in skeletal muscle tissue. In addition, it has been shown in both experimental animals (Davisson et al. 1993) and humans (Saenz de Tejada et al. 1989) that neural nonadrenergic noncholinergic vasodilation is mediated by NO release. These observations could be consistent with the concept that insulin stimulates NO release by activation of neural noncholinergic nonadrenergic pathways. This interpretation is strengthened by the observation that systemic insulin/glucose infusion consistently has been found to evoke large increases in muscle blood flow (Laakso et al. 1990a; Anderson et al. 1991; Bennet et al. 1990; Richter et al. 1989; Vollenweider et al. 1993, 1994), whereas studies using local, intraarterial insulin infusion into human forearm tissue have shown divergent results (Andres et al. 1961; Chrisholm et al. 1975; Ferrannini et al. 1988; Gelfand and Barret 1987; Natali et al. 1990). For example, the study by Natali and colleagues (1990) showed that local intraarterial insulin infusion which markedly stimulated carbohydrate metabolism, failed to stimulate muscle blood flow, whereas systemic insulin infusion resulting in similar plasma insulin concentrations as the ones obtained by Natali et al. consistently has been shown to stimulate muscle blood flow.

We do not know yet the exact mechanism by which insulin stimulates NO release, nor how it may be altered in insulin resistant states. What these recent data show, however, is that stimulation of NO release plays a major role in the mediation of insulin-induced vasodilation, and that suppression of such vasodilation leads to increases in arterial pressure during acute physiologic hyperinsulinemia. A defect in insulin-induced stimulation of NO-release could be one of the factors that contributes to altered vascular function (Saenz de Tejada et al. 1989; Vollenweider et al. 1994; Laakso et al. 1992a; Baron et al. 1993) and hypertension (Modan et al. 1985) in insulin-resistant states.

Effects of Induction of Acute Insulin Resistance in Lean Subjects on Insulin-Induced Sympathetic and Vasodilator Responses

In obese insulin-resistant subjects, insulin's ability to stimulate sympathetic activity and muscle blood flow is impaired (Laakso et al. 1990a; Vollenweider et al. 1994) but the underlying mechanism is unknown. More specifically, it is not known whether in obese subjects, resistance to insulin's stimulatory effects on muscle glucose uptake contributes to attenuated vasodilation and sympathetic activation.

To examine whether in lean subjects, attenuation of insulin's metabolic effects alters the sympathetic and vasodilatory responses evoked by euglycemic hyperinsulinemia Vollenweider and colleagues (1995) employed free fatty acid infusion. In lean subjects, lipid infusion by its effect on the glucose free fatty acid cycle, markedly attenuates insulin-induced stimulation of glucose uptake and oxidation (Randle et al. 1963); and there is evidence that in obese subjects, substrate competition between lipids and glucose contributes to insulin resistance (Felber et al. 1993). The main finding of these studies was that free fatty acid infusion which attenuated insulin-induced stimulation in these lean subjects by roughly 40% did not have any detectable effect on insulin-induced stimulation of sympathetic nerve discharge and blood flow in skeletal muscle tissue. These observations indicate that induction of resistance to insulin's stimulatory effects on carbohydrate metabolism does not attenuate muscle blood flow and sympathetic responses to insulin infusion. They suggest that the impairment in sympathetic and vasodilatory responsiveness to insulin in obese subjects is not related primarily to the impairment in insulin-induced stimulation of muscle carbohydrate metabolism.

Vascular and Sympathetic Actions of Insulin in Insulin-Resistant States

Effects of Obesity on Insulin-Induced Sympathetic Activation and Vasodilation in Skeletal Muscle

In obese insulin-resistant subjects, insulin looses its ability to stimulate muscle blood flow (Laakso et al. 1990a), however the underlying defect leading to such an impairment remains unknown. We have shown earlier in this chapter that in lean subjects, insulin-induced stimulation of muscle blood flow consistently is accompanied by a marked stimulation of sympathetic neural outflow to the same vascular bed. Theoretically, such sympathetic activation could modulate muscle blood flow in two (directionally opposite) ways. First, this sympathetic activity is thought to be, at least in part, sympathetic vasoconstrictor because it is accompanied by stimulation of norepinephrine release (Lembo et al. 1992). Alternatively, such sympathetic activation, by stimulation of neural sympathetic vasodilator pathways, could contribute to vasodilation in skeletal muscle. Data in experimental animals regarding effects of insulin resistance and obesity on insulin-induced sympathetic responses are conflicting. On the one hand, there is evidence in insulin resistant rats that sympathetic nerve responses to insulin are markedly augmented (Morgan et al. 1991). On the other hand, in models of genetically inherited obesity in rodents diet-induced sympathetic responsiveness is impaired (Holt et al. 1983). Thus, while there is evidence in experimental animals that obesity may alter sympathetic nerve responsiveness, this evidence is conflicting, and no data exist in humans.

To provide such data, in a recent study we have compared muscle sympathetic nerve and blood flow responses to insulin/glucose infusion in obese (body mass index >30 kg/m^2) and lean (body mass index <25 kg/m^2) subjects (Vollenweider et al. 1994). The principal new findings were twofold: in the fasting state, obese subjects had 2.2 times higher rates of sympathetic nerve discharge than lean subjects; and euglycemic hyperinsulinemia which more than doubled sympathetic activity and increased calf blood flow by roughly 30% in lean subjects, had only a barely detectable sympathetic and vasodilatory effect in obese subjects (Fig. 8). This latter observation is not related to a non-specific impairment of sympathetic responsiveness in overweight subjects because reflex sympathetic responses during a Valsalva maneuver and during immersion of the hand in ice water were preserved, indicating that in obese subjects efferent sympathetic pathways could respond appropriately to baroreceptor deactivation and stimulation of cutaneous afferents. The impaired sympathetic responsiveness to insulin infusion in obese subjects also cannot be explained on the basis of resistance to insulin stimulation of glucose uptake alone, because in lean subjects, attenuated stimulation of glucose uptake during low dose insulin infusion did not result in attenuated stimulation of sympathetic nerve discharge and muscle blood flow.

We therefore considered an alternative possibility to explain the impaired sympathetic responsiveness to insulin infusion in obese subjects. We found that obese subjects had a roughly twofold higher rate of sympathetic nerve discharge at baseline than lean subjects, a rate that tended to be still higher than the rate observed in lean subjects at the end of a 2-h insulin infusion. This observation raises the possibility the high basal rates of sympathetic discharge in obese subjects may lie close to maximal increase in sympathetic activity that can be obtained by acute elevation of insulinemia to high physiological concentrations in humans. This hypothesis is supported by the results of the low dose insulin infusion studies in lean subjects (Fig. 3); these data suggest that during insulin infusion at a rate of 1 mU/kg per minute (and at a rate of 0.5 mU/kg per minute) sympathetic responses do not fall an the steep slope, but already fall on the flat part of the curve which relates plasma insulin concentration to efferent muscle sympathetic nerve discharge. Thus, two possible interpretations can be considered to explain the impaired sympathetic responsiveness in obese subjects.

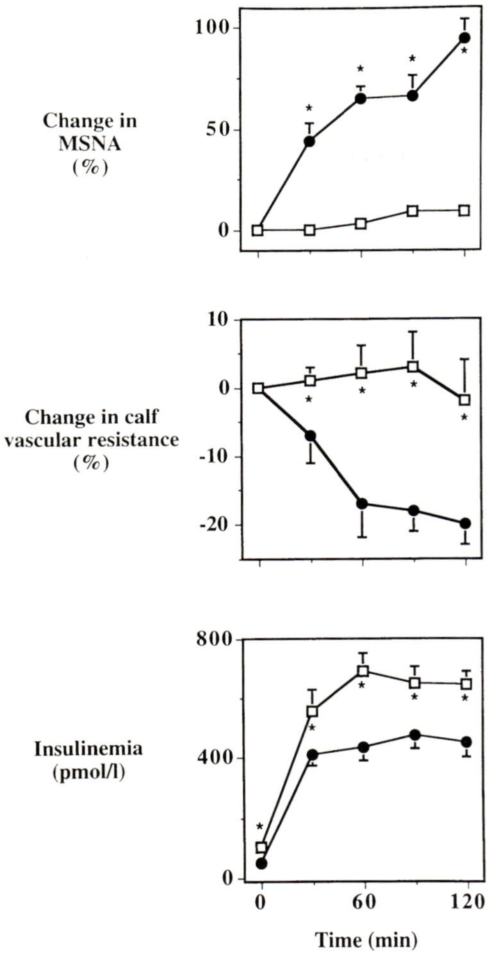

Fig. 8. Effects of a 2-h insulin infusion (1 μU/kg per minute, euglycemic hyperinsulinemic clamp) on muscle sympathetic nerve activity (MSNA), calf vascular resistance, and plasma insulin concentration in lean (*closed circles*) and obese (*open squares*) subjects. Data represent means ± SE for eight subjects in each group. * $p < 0.05$ lean vs obese subjects. (From Vollweider et al. 1994)

First, the attenuated responsiveness could be related to a resistance to insulin's sympathoexcitatory effects and thereby represent another feature of obesity-induced insulin resistance. Alternatively, the high rate of baseline sympathetic nerve discharge in obese subjects, which could possibly be related to chronic hyperinsulinemia, may lie close to the maximal insulin-induced sympathetic activation in skeletal muscle and thus may preclude the demonstration of any further increase during acute elevation of plasma insulin concentrations to high physiological levels. With regard to the former, there is evidence for an impairment in sympathetic responsiveness in genetically obese animals (Holt et al. 1983), and recent findings in humans indicate that short term dexamethasone administration, possibly by interfering with insulin's central neural actions, impairs insulin's ability to stimulate sympathetic activity and blood flow in skeletal muscle (Scherrer et al. 1993). On the other hand, preserved sensitivity to insulin's sympathoexcitatory effects in chronically hyperinsulinemic subjects may lead to sustained sympathetic activation and could represent one potential mechanism contributing to the increased incidence of cardiovascular complications in overweight subjects (Modan and Halkin 1991).

In conclusion, these data in humans show that obesity is associated with profound alterations in sympathetic nerve activity characterized by increased fasting activity and a specific attenuation of sympathetic responsiveness to hyperinsulinemia.

Body Fat and Sympathetic Nerve Activity in Healthy Subjects

Obesity, a major health problem in industrialized societies, is associated with a high incidence of cardiovascular complications such as hypertension, ischemic heart disease and stroke (Modan and Halkin 1991; Dustan 1991; Levy et al. 1946; Gordon and Kannel 1976). However, the underlying mechanism relating obesity and these cardiovascular events is not clear. In experimental animal models overfeeding stimulates sympathetic activity (Young and Landsberg 1977; Young et al. 1982), and because overfeeding is associated with insulin resistance and hyperinsulinemia, the question whether these two factors may contribute to sympathetic activation is an intensively researched but unresolved issue (Modan and Halkin 1991; Dustan 1991; Daly and Landsberg 1991). Sympathetic activation could conceivably contribute to increased systemic vascular resistance and hypertension due to activation of alpha-adrenergic vasoconstriction, and stimulation of the renin-angiotensin-system (Shepherd 1990). Such vasoconstriction could be reinforced by sympathetically mediated trophic effects on the vasculature (Bevan 1984). Furthermore, sympathetic activation has been shown to promote atherosclerosis, and could trigger acute cardiovascular events by increasing platelet number and aggregability (Sloan et al. 1989; Clayton and Cross 1963). Surprisingly however, studies in humans examining the relation between sympathetic activity and body weight have produced conflicting results, and shown either decreased, normal, or augmented sympathetic activity in overweight subjects (Young and MacDonald 1992). While many factors may explain such divergent findings in humans, the possibility that regional heterogeneity in sympathetic activity may underlie some of the variability in lean/obese comparisons has received little attention. Microneurography allows direct recording of postganglionic sympathetic nerve action potentials targeted at the skeletal muscle vascular bed (Vallbo et al. 1979). We therefore recently have measured sympathetic nerve activity in skeletal muscle in a group of subjects covering a broad spectrum of percentages of body fat (Scherrer et al. 1994a). To examine potential functional consequences of sympathetic nerve discharge, we simultaneously measured calf vascular resistance, arterial pressure and energy expenditure.

We found that sympathetic nerve activity is closely and directly related to percent body fat and body mass index (Fig. 9). Indeed, our results indicate that in healthy subjects body fat is a major determinant of sympathetic activity and accounts for roughly 50% of the interindividual variation in the rate of sympathetic nerve firing in skeletal muscle. Using simultaneous measurements of vascular resistance to examine potential functional consequences of such overweight associated sympathetic activation we found that the rate of sympathetic nerve discharge to calf vasculature was directly correlated with calf vascular resistance. This increase in peripheral vascular

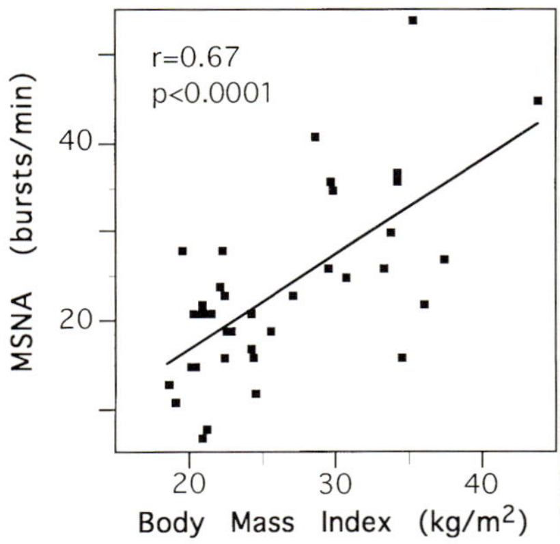

Fig. 9. Relationship between body mass index and the rate of sympathetic nerve discharge to skeletal muscle vasculature. *MSNA*, muscle sympathetic nerve activity. (From Scherrer et al. 1994a)

resistance could represent one of the factors contributing to the positive correlation found between body mass index and arterial pressure found in these subjects.

The sympathetic nervous system not only is involved in cardiovascular control but also in the regulation of energy expenditure, and sympathetic activation has been shown to stimulate energy expenditure (Acheson et al. 1983). Recent work in humans suggests that skeletal muscle metabolism may account for part of the variance in basal metabolic rate among individuals (Zurlo et al. 1990). In our study, we did not find a direct relationship between sympathetic discharge to skeletal muscle and the deviation from predicted energy expenditure as determined by indirect calorimetry. This conclusion differs from that of a recent report by Spraul and colleagues (1993), indicating that there is not only, and in agreement with the present data, a positive correlation between body fat and sympathetic nerve activity, but also a significant relationship between muscle sympathetic nerve activity and energy expenditure. However, our findings are in agreement with recent data of the same group, indicating that there is no relationship between sympathetic activity and resting metabolic rate in men (Christin et al. 1993). Further studies are needed to examine whether such discharge is a determinant of resting muscle metabolism.

Although the mechanism by which overweight could lead to sympathetic activation remains unknown at least two possibilities might be considered. In rats, the sympathetic activation evoked by overfeeding is associated with insulin resistance and hyperinsulinemia (Hwang et al. 1987). In both rats (Morgan et al. 1993) and humans (Anderson et al. 1991; Vollenweider et al. 1993), acute hyperinsulinemia triggers sympathetic activation in skeletal muscle tissue. Two observations in our study could be consistent with the hypothesis that hyperinsulinemia may be one of the factors contributing to sympathetic activation in overweight humans. First, we found a positive correlation between fasting plasma insulin concentration and the rate of sympathetic firing. Second, when we used low dose insulin infusion in lean subjects to increase their plasma insulin concentrations to levels similar to those observed in fasting overweight subjects, we observed a marked increase in sympathetic activity (Fig. 3).

Although, undoubtedly many factors contribute to the increased incidence of cardiovascular complications in overweight subjects, sympathetic activation could be one important mechanism and either trigger acute events (Muller et al. 1989) or – possibly in conjunction with an impairment in insulin-induced vasodilation (Laakso et al. 1990a; Vollenweider et al. 1994) – contribute to sustained elevation of arterial pressure. With regard to the latter, recent data suggest that in overweight subjects, weight loss-induced reductions in arterial pressure are associated with decreases in both sympathetic vasoconstrictor outflow to skeletal muscle (Andersson et al. 1991), and peripheral vascular resistance (Jacobs et al. 1993). Finally, sympathetic activation could also represent a potential link between essential hypertension and insulin resistance in lean subjects (Ferrannini et al. 1987; Baron et al. 1993).

Conclusions

Insulin, apart from its effects on intermediary metabolism has effects on the cardiovascular system that until recently have received little attention (Shepherd and Vanhoutte 1979). In this review, we have provided strong evidence that in lean subjects, insulin (even at very low doses) exerts important effects on the cardiovascular system. These effects which are mediated both by the sympathetic nervous system and the L-arginine nitric oxide pathway appear to play a role in the regulation of blood pressure. The hypothesis that insulin's vascular effects also play an important role in its main action, namely the promotion of glucose disposal in skeletal muscle, has recently been challenged and this issue awaits further study.

Obese, insulin-resistant subjects are characterized by profound alterations in both sympathetic and vascular responsiveness to acute elevation of plasma insulin concentrations. Indeed, insulin resistance in obese subjects is associated with a specific impairment in vasodilator and sympathetic neural responsiveness to acute hyperinsulinemia in skeletal muscle tissue. Moreover, overweight is associated with marked and sustained sympathetic activation which could be related to chronic hyperinsulinemia. Such sympathetic activation – possibly in conjunction with an impairment in L-arginine nitric oxide dependent vasodilation – could represent one important mechanism that contributes to the increased incidence of both acute and chronic cardiovascular complications in such patients.

Abstract

Insulin exerts effects on the cardiovascular system that may play a role in the regulation of blood pressure and, by boosting its own delivery to target tissues, also appears to play an integral part in its main action, the promotion of glucose disposal. Surprisingly, until recently the cardiovascular effects of insulin have received little attention.

In this review we provide strong evidence that in lean healthy subjects, even slight elevations in plasma insulin concentrations exert marked effects on the cardiovascular system, and that these effects are mediated by both the sympathetic nervous system and the L-arginine nitric oxide pathway. Insulin's sympathoexcitatory effects which appear to be centrally mediated, and its vasodilatory effects in the skeletal muscle vasculature which involve stimulation of nitric oxide release play a role in the regulation of blood pressure during acute physiologic hyperinsulinemia.

Obese insulin-resistant subjects are characterized by impaired vasodilator responsiveness to insulin and by sustained sympathetic activation. Sustained sympathetic activation - possibly in conjunction with impaired L-arginine nitric dependent vasodilation - could participate in triggering cardiovascular events in overweight subjects.

Acknowledgments. I would like to thank Drs. Peter Vollenweider, Denis Randin, Laurent Vollenweider, Reza Owlya, and Luc Tappy for their invaluable contributions to these studies; Dr. Eric Jéquier for his continued support; and Dr. Pascal Nicod for many stimulating discussions, his critical review of this manuscript and his superb support. This work was supported in part by grants from the Swiss National Science Foundation (32.28668.90 and 32-36280.92), the Emma Muschamp Foundation, the Sandoz Foundation, the International Olympic Committee, Nestec SA Vevey, Switzerland, the Max Cloëtta Foundation, the Jubiläumsstiftung der Schweizerischen Rentenanstalt, and the Placide Nicod Foundation.

References

Abramson DI, Schkloven N, Margolis MN, Mirsky IA (1939) Influence of massive doses of insulin on peripheral blood flow in man. Am J Physiol 128:124–132

Acheson K, Jéquier E, Wahren J (1983) Influence of beta-adrenergic blockade on glucose-induced thermogenesis in man. J Clin Invest 72:981–986

Anderson EA, Hoffman RP, Balon TW, Sinkey CA, Mark AL (1991) Hyperinsulinemia produces both sympathetic neural activation and vasodilation in normal humans. J Clin Invest 87:2246–2252

Andersson B, Elam M, Wallin BG, Björntop P, Andersson OK (1991) Effect of energy-restricted diet on sympathetic muscle nerve activity in obese woman. Hypertension 18:783–789

Andres M, Baltzan MA, Cader G, Zierler KL (1961) Effect of insulin on carbohydrate metabolism and on potassium in the forearm in man. J Clin Invest 41:108–115

Baron AD (1993) Cardiovascular actions of insulin in humans. Implications for insulin sensitivity and vascular tone. Baillieres Clin Endocrinol Metab 7:961–987

Baron AD, Brechtel-Hook G, Johnson A, Hardin D (1993) Skeletal muscle blood flow: a possible link between insulin resistance and blood pressure. Hypertension 21:129–135

Bennet WM, Connacher AA, Scrimgeour CM, Jung JT, Rennie MJ (1990) Euglycemic hyperinsulinemia augments amino acid uptake by human leg tissues during hyperaminoacidemia. Am J Physiol 259:E185–E194

Berne C, Fagius J, Pollare T, Hjemdahl P (1992) The sympathetic response to euglycaemic hyperinsulinaemia: evidence from microelectrode nerve recordings in healthy subjects. Diabetologia 35: 873–879

Bevan RD (1984) Trophic effects of peripheral adrenergic nerves on vascular structure. Hypertension 6 [Suppl III]:III19–III26

Bonora E, Zavaroni I, Alpi O, Pezzarossa A, Bruschi F, Dall'Aglio E, Guerrra L, Coscelli C, Butturini U (1987) Relationship between blood pressure and plasma insulin in non-obese and obese non-diabetic subjects. Diabetologia 30:719–723

Brown MR, Fisher LA (1985) Corticotropin-releasing factor: effects on the autonomic nervous system and visceral systems. Fed Proc 44:243–248

Calver A, Collier J, Vallance P (1992) Inhibition and stimulation of nitric oxide synthesis in the human forearm arterial bed of patients with insulin-dependent diabetes. J Clin Invest 90:2548–2554

Chrisholm DJ, Klassen GA, Dupre J, Pozefski T (1975) Interaction of secretin and insulin on human forearm metabolism. Eur J Clin Invest 5:487–494
Christin L, O'Connell M, Bogardus C, Danforth E, Ravussin E (1993) Norepinephrine turnover and energy expenditure in Pima Indian and white men. Metabolism 42:723–729
Clayton S, Cross M (1963) The aggregation of blood platelets by catecholamines and by thrombin. J Physiol (Lond) 169:82P
Daly PA, Landsberg L (1991) Hypertension in obesity and NIDDM. Diabetes Care 14:240–248
Davisson RL, Hoque AM, Johnson AK, Lewis SJ (1993) Sympathetic vasodilator neurons innervating rat hindlimb vasculature may release nitrosyl factors. Hypertension 22:413
DeFronzo RA, Tobin JD, Andres R (1979) Glucose clamp technique: a method for quantifying insulin secretion and resistance. Am J Physiol 237:E214–E223
DiSalvo RJ, Bloom WL, Brust AA, Ferguson RW, Ferris EB (1956) A comparison of the metabolic and circulatory effects of epinephrine, norepinephrine and insulin hypoglycemia with observations on the influence of autonomic blocking agents. J Clin Invest 35:568–577
Dustan HP (1991) Obesity and hypertension. Diabetes Care 14:488–504
Felber JP, Haesler E, Jéquier E (1993) Metabolic origin of insulin resistance in obesity with and without Type 2 (non-insulin-dependent) diabetes mellitus. Diabetologia 36:1221–1229
Ferrannini E, Buzzigogli RD, Bonadonna R, Giorico MA, Oleggini M, Graziadei L, Pedrinelli R, Brandi L, Bevilaqua S (1987) Insulin resistance in essential hypertension. N Engl J Med 317:350–357
Ferrannini E, Taddei S, Santoro D, Natali A, Boni C, Del Chiaro D, Buzzigoli G (1988) Independent stimulation of glucose metabolism and Na-K exchange by insulin in the human forearm. Am J Physiol 255:E953–E958
Gelfand RA, Barret EJ (1987) Effect of physiological hyperinsulinemia on skeletal muscle protein synthesis and breakdown in man. J Clin Invest 80:1–6
Ginsburg J, Paton A (1956) Effects of insulin after adrenalectomy. Lancet 271:491–494
Gordon T, Kannel WB (1976) Obesity and cardiovascular disease: the Framingham study. Clin Endocrinol Metab 5:367–375
Gosink PD, Chauvez M, Green PK, Woods SC, Figlewicz DP, Wilkinson CW, Baskin DG, Schwartz MW (1992) Central insulin administration lowers body weight via a glucocorticoid sensitive mechanism. Clin Res 40:55A
Gruber KA, Callahan MF (1989) ACTH-(4–10) through gamma-MSH: evidence for a new class of central autonomic system-regulating peptides. Am J Physiol 257:R681–R694
Hohnloser J, Von Werder K, Müller OA (1989) Acute dexamethasone suppression of ACTH secretion stimulated by human corticotrophin releasing hormone. Clin Endocrinol (Oxf) 31:175–184
Holt S, York DA, Fitzsimons JTR (1983) The effect of corticosterone, cold exposure and overfeeding with sucrose on brown adipose tissue of obese Zucker rats (fa/fa). Biochem J 214:215–223
Hwang IS, Ho H, Hoffman BB, Reaven GM (1987) Fructose-induced insulin resistance and hypertension in rats. Hypertension 14:117–119
Jacobs DB, Sowers JR, Hmeidan A, Niyogi T, Simpson L, Standley PR (1993) Effects of weight reduction on cellular cation metabolism and vascular resistance. Hypertension 21:308–314
Kurosawa M, Sato A, Swenson R, Takahashi Y (1985) Sympatho-adrenal medullary functions in response to intracerebroventricular injected corticotropin-releasing factor in anesthetized rats. Brain Res 367:250–257
Laakso M , Edelman SV, Brechtel G, Baron AD (1990a) Decreased effect of insulin to stimulate muscle blood flow in obese man. A novel mechanism for insulin resistance. J Clin Invest 85:1844–1852
Laakso M, Edelman SV, Olefsky JM, Brechtel G, Wallace P, Baron AD (1990b) Kinetics of in vivo muscle insulin-mediated glucose uptake in human obesity. Diabetes 39:965–974
Laakso M, Edelman SV, Brechtel G, Baron AD (1992a) Impaired insulin-mediated skeletal muscle blood flow in patients with NIDDM. Diabetes 41:1076–1083
Laakso M, Edelman SV, Brechtel G, Baron AD (1992b) Effects of epinephrine on insulin-mediated glucose uptake in whole body and leg muscle in humans: role of blood flow. Am J Physiol 263: E199–E204
Landsberg L (1986) Diet, obesity and hypertension: a hypothesis involving insulin, the sympathetic nervous system and adaptive thermogenesis. Q J Med 236:1081–1090
Lembo G, Napoli R, Capaldo B, Rendina V, Iaccarino G, Volpe M, Trimarco B, Sacca L (1992) Abnormal sympathetic overactivity evoked by insulin in skeletal muscle of patients with essential hypertension. J Clin Invest 90:24–29

Levy RL, White PD, Stroud WD, Hillman CC (1946) Overweight: prognostic significance in relation to hypertension and cardiovascular disease. JAMA 131:951–953

Liang C-S, Doherty JU, Faillace R, Maekawa K, Arnold S, Gavras H, Hood WB (1982) Insulin infusion in conscious dogs: effects on systemic and coronary hemodynamics, regional blood flows, and plasma catecholamines. J Clin Invest 69:1321–1336

Lillioja S, Young AA, Culter CL, Ivy JL, Abbot WGH, Zawadski JK, Yki-Järvinen H, Christin L, Secomb TW, Bogardus C (1987) Skeletal muscle capillary density and fiber type are possible determinants of in vivo insulin resistance in man. J Clin Invest 1987:415–425

Lucas CP, Estigarribia JA, Darga LL, Reaven GM (1985) Insulin and blood pressure in obesity. Hypertension 7:702–706

Margolis RU, Altszuler N (1967) Insulin in the cerebrovascular fluid. Nature 215:1375–1376

Modan M, Halkin H (1991) Hyperinsulinemia or increased sympathetic drive as links for obesity and hypertension. Diabetes Care 14:470–487

Modan M, Halkin H, Almog S, Lusky A, Eshkol A, Shefi M, Shitrit A, Fuchs Z (1985) Hyperinsulinemia. A link between hypertension, obesity, and glucose intolerance. J Clin Invest 75:809–817

Morgan DA, Balon TW, Mark AL (1991) High fructose diet promotes exaggerated sympathetic and pressor responses to hyperinsulinemia in rats. FASEB J 5:A1497

Morgan DA, Balon TW, Ginsberg BH, Mark AL (1993) Nonuniform regional sympathetic nerve responses to hyperinsulinemia in rats. Am J Physiol 264:R423–R427

Muller JE, Tofler GH, Stone PH (1989) Circadian variation and triggers of onset of acute cardiovascular disease. Circulation 80:267–275

Natali A, Buzzigoli G, Taddei S, Santoro D, Cerri M, Pedrinelli R, Ferrannini E (1990) Effects of insulin on hemodynamics and metabolism in the human forearm. Diabetes 39:490–500

Overton JM, Fischer LA (1989) Central nervous system actions of corticotropin-releasing factor on cardiovascular function in the absence of locomotor activity. Regul Pept 25:315–324

Pereda SA, Eckstein JW, Abboud FM (1962) Cardiovascular responses to insulin in the absence of hypoglycemia. Am J Physiol 202:249–252

Randin D, Vollenweider P, Tappy L, Jéquier E, Nicod P, Scherrer U (1994) Effects of adrenergic and cholinergic blockade on insulin-induced stimulation of calf blood flow in humans. Am J Physiol 266:R809–R816

Randin D, Vollenweider P, Tappy L, Jéquier E, Nicod P, Scherrer U (1995) Suppression of alcohol-induced sympathetic activation and elevation of arterial pressure by dexamethasone. N Engl J Med 332 (in press)

Randle PJ, Garland PB, Hales CN, Newsholme EA (1963) The glucose fatty-acid cycle: its role in insulin sensitivity and metabolic disturbances of diabetes mellitus. Lancet i:785–789

Reaven GM (1988) Role of insulin resistance in human disease. Diabetes 37:1595–1607

Richter EA, Mikines KJ, Galbo H, Kiens B (1989) Effect of exercise on insulin action in human skeletal muscle. J Appl Physiol 66:876–885

Rowe JW, Young JB, Minaker KL, Stevens AL, Pallota J, Landsberg L (1981) Effects of insulin and glucose infusions on sympathetic nervous system activity in normal man. Diabetes 30:219–225

Saenz de Tejada I, Goldstein I, Azadzoi K, Krane RJ, Cohen RA (1989) Impaired neurogenic and endothelium-mediated relaxation of penile smooth muscle from diabetic men with impotence. N Engl J Med 320:1025–1030

Sauter A, Goldstein M, Engel J, Ueta K (1983) Effect of insulin on central catecholamines. Brain Res 260:330–333

Scherrer U, Vissing SF, Morgan BJ, Rollins JA, Tindall RSA, Ring S, Hanson P, Mohanty PK, Victor RG (1990) Cyclosporine-induced sympathetic activation and hypertension after heart transplantation. N Engl J Med 323:693–699

Scherrer U, Vollenweider P, Randin D, Jéquier E, Nicod P, Tappy L (1993) Suppression of insulin induced sympathetic activation and vasodilation by dexamethasone in humans. Circulation 88:388–394

Scherrer U, Randin D, Tappy L, Vollenweider P, Jéquier E, Nicod P (1994a) Body fat and sympathetic nerve activity in healthy subjects. Circulation 89:2634–2640

Scherrer U, Randin D, Vollenweider P, Vollenweider L, Nicod P (1994b) Nitric oxide release accounts for insulin's vascular effects in humans. J Clin Invest 94:2511–2515

Schwartz MW, Figlewicz DP, Baskin DG, Woods SC, Porte DJ (1992) Insulin in the brain: a hormonal regulator of energy balance. Endocr Rev 13:387–414

Scott NA, Webb V, Boublik JH, Rivier J, Brown MR (1989) The cardiovascular actions of centrally administered neuropeptide Y. Regul Pept 25:247–258

Shen DC, Shieh SM, Fuh MT, Wu DA, Chen YDI, Reaven GM (1988) Resistance to insulin-stimulated glucose uptake in patients with hypertension. J Clin Endocrinol Metab 66:580–583

Shepherd JT (1990) Increased systemic vascular resistance and primary hypertension: the expanding complexity. J Hypertens 8 [Suppl 7]:S15–S27

Shepherd JT, Vanhoutte PM (1979) Interactions between cardiovascular and endocrine control. In: Shepherd JT, Vanhoutte PM (eds) The human cardiovascular system. Facts and concepts. Raven, New York, p 153

Sloan JA, Hooper M, Izzo JL Jr (1989) Effects of circulating norepinephrine on platelets, leukocyte and RBC counts by alpha1-adrenergic stimulation. Am J Cardiol 63:1140–1142

Spraul M, Ravussin E, Fontvieille AM, Rising R, Larson DE, Anderson EA (1993) Reduced sympathetic nerve activity. A potential mechanism predisposing to body weight gain. J Clin Invest 92:1730–1735

Steinberg HO, Brechtel G, Johnson A, Fineberg N, Baron AD (1994) Insulin-mediated skeletal muscle vasodilation is nitric oxide dependent. A novel action of insulin to increase nitric oxide release. J Clin Invest 94:1172–1179

Tagawa T, Imaizumi T, Endo T, Shiramoto M, Hirooka Y, Ando S, Takeshita A (1993) Vasodilatory effect of arginine vasopressin is mediated by nitric oxide in human forearm vessels. J Clin Invest 92:1483–1490

Takeshita AT, Mark AL (1980) Decreased vasodilator capacity of forearm resistance vessels in borderline hypertension. Hypertension 2:610–616

Vallance P, Collier J, Moncada S (1989) Effects of endothelium-derived nitric oxide on peripheral arteriolar tone in man. Lancet ii:997–1000

Vallbo AB, Hagbarth KE, Torebjörk HE, Wallin BG (1979) Somatosensory, proprioceptive, and sympathetic activity in human peripheral nerves. Physiol Rev 59:919–957

Van Houten M, Posner BI (1979) Insulin binds to brain blood vessels. Nature 282:623–625

Victor RG, Leimbach WN, Seals DR, Wallin BG, Mark AL (1987) Effects of the cold pressure test on muscle sympathetic nerve activity in humans. Hypertension 9:429–436

Vollenweider P, Tappy L, Randin D, Schneiter P, Jéquier E, Nicod P, Scherrer U (1993) Differential effects of hyperinsulinemia and carbohydrate metabolism on sympathetic nerve activity and muscle blood flow in humans. J Clin Invest 92:147–154

Vollenweider P, Randin D, Tappy L, Jéquier E, Nicod P, Scherrer U (1994) Impaired insulin-induced sympathetic neural activation and vasodilation in skeletal muscle in obese humans. J Clin Invest 93:2365–2371

Vollenweider L, Tappy L, Owlya R, Jéquier E, Nicod P, Scherrer U (1995) Insulin-induced sympathetic activation and vasodilation in skeletal muscle. Effects of insulin resistance in lean subjects. Diabetes 44 (in press)

Yang YI, Hope ID, Ader M, Bergman RN (1989) Insulin transport across capillaries is rate limiting for insulin action in dogs. J Clin Invest 84:1620–1628

Young JB, Landsberg L (1977) Stimulation of the sympathetic nervous system during sucrose feeding. Nature 269:615–617

Young JB, Saville E, Rothwell NJ, Stock MJ, Landsberg L (1982) Effect of diet and cold exposure on norepinephrine turnover in brown adipose tissue in the rat. J Clin Invest 69:1061–1071

Young JB, Macdonald IA (1992) Sympathoadrenal activity in human obesity: heterogeneity of findings since 1980. Int J Obes 16:959–967

Zurlo F, Larson K, Bogardus C, Ravussin E (1990) Skeletal muscle metabolism is a major determinant of resting energy expenditure. J Clin Invest 86:1423–1427

Endothelin Receptor Antagonists as New Tools to Inhibit Endothelin induced Vasoconstriction in Humans: Comparison with Calcium Channel Blockers

R. R. Wenzel and T. F. Lüscher

Introduction

Diseases of the cardiovascular system account for the majority of morbidity and mortality in Western countries. Most forms of cardiovascular disease are associated with impaired vascular function in the cerebral, coronary, renal or peripheral circulation. Enhanced intermittent and/or persistent vasoconstriction is one of the most important features of several forms of cardiovascular disease [1]. Indeed, hypertension is characterized by a persistent increase in vascular tone of resistance arteries. In coronary spasm, migraine and Raynaud's phenomenon, focal spasms of larger arteries are responsible for the clinical symptoms. Enhanced coronary vasoconstriction also importantly determines the angina threshold in patients with atherosclerotic coronary artery disease. Similarly, in low-tension glaucoma, increased vascular tone has been accounted for vision field defects and loss of sight [2]. Cerebral vasospasm after subarachnoid hemorrhage is an important complication associated with high morbidity and mortality in patients [3].

In all these clinical syndromes, the mechanisms of enhanced vasoconstriction of either large conduit arteries or the resistance circulation is still only poorly understood. An increased activity of the sympathetic nervous system may contribute in certain disease states, although this is also controversial. Circulating hormones are unlikely to explain focal vasospasm characteristic of variant angina, Raynaud's phenomenon or migraine. Hence, the interest recently has focused on local vascular regulatory systems which may fail very locally and explain why certain arterial segments but not others are prone to the disease.

An important local vascular regulator is the recently discovered 21-amino acid peptide endothelin [4–6]. Endothelin exists in three isoforms, i.e. endothelin-1, -2 and -3. In the blood vessel wall, only endothelial cells produce the peptide and exclusively endothelin-1, except in disease states such as atherosclerosis, where smooth muscle cells may also contribute. Most of the endothelin produced in the endothelium is released abluminally to interact with vascular smooth muscle, while only about a third of it reaches the circulating blood where it can be measured by radioimmunoassay techniques [7]. Three endothelin receptors have been cloned, i.e. ET_A, ET_B and ET_C receptors. While the role of ET_C receptors remains uncertain [8, 9], it is now clear that vascular smooth muscle cells express both ET_A and ET_B receptors linked to vasoconstriction and proliferation of vascular smooth muscle cells [10], while endothelial cells exclusively express ET_B receptors linked to the formation of nitric oxide and prostacyclin (Fig. 1). Hence, with intraluminal administration of endothelin-1, a tran-

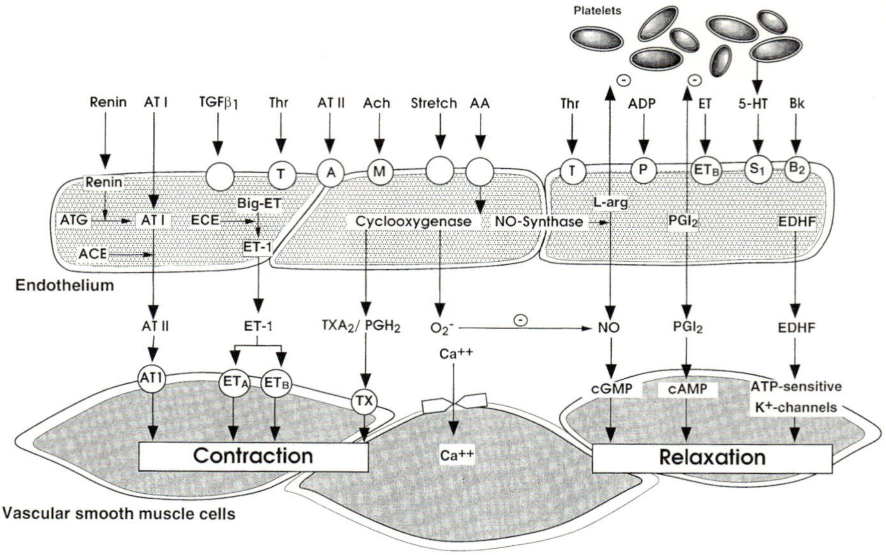

Fig. 1. The vascular endothelium produces a variety of substances interacting either luminally with blood cells such as platelets or abluminally with vascular smooth muscle cells. Endothelin (*ET*) is the most potent vasoconstrictor known and is produced in the endothelium from its precursor big-ET via the endothelin-converting enzyme (*ECE*). It is mainly released abluminally towards vascular smooth muscle cells and interacts with specific receptors (*open circles*). On the endothelium, ET activates receptors linked to nitric oxide (*NO*) and prostacyclin (*PGI_2*) formation. *AT*, angiotensin; *TGF*, transforming growth factor; *Ach*, acetylcholine; *ADP*, Adenosine diphosphate; *5-HT*, 5-hydroxytryptamine; *ACE*, angiotensin-converting enzyme; *TXA*, thromboxane A; *PGH*, prostaglandin H; *EDHF*, endothelium-derived hyperpolarizing factor; *cGMP*, cyclic guanosine monophosphate; *cAMP*, cyclic adenosine monophosphate; *ATP*, adenosine triphosphate; *L-arg*, L-arginine; *AA*, arachidonic acid; *Bk*, bradykinin; *B_2*, bradykinin receptor; *S_1*, serotonin receptor; *P*, purinergic receptor; *T*, thrombin; *M*, muscarinergic receptor; *A*, angiotensin receptor

sient vasodilation due to activation of nitric oxide and prostacyclin is observed followed by profound and sustained vasoconstriction which is mediated by ET_A receptor activation and at least in certain blood vessels also by ET_B receptors [11]. With higher concentrations of endothelin only vasoconstriction is observed, although even then inhibition of a production of endothelium-derived relaxing factors enhances the response to the peptide [12].

Endothelin and Cardiovascular Disease

The production of endothelin is increased in various forms of cardiovascular disease (Fig. 2) [13, 14]. Indeed, both atherosclerosis and inflammatory vascular disease such as Takayasu's disease are associated with increased plasma endothelin levels. In contrast, although some authors also reported increased plasma endothelin levels in Raynaud's disease [15], we were unable to confirm these findings.

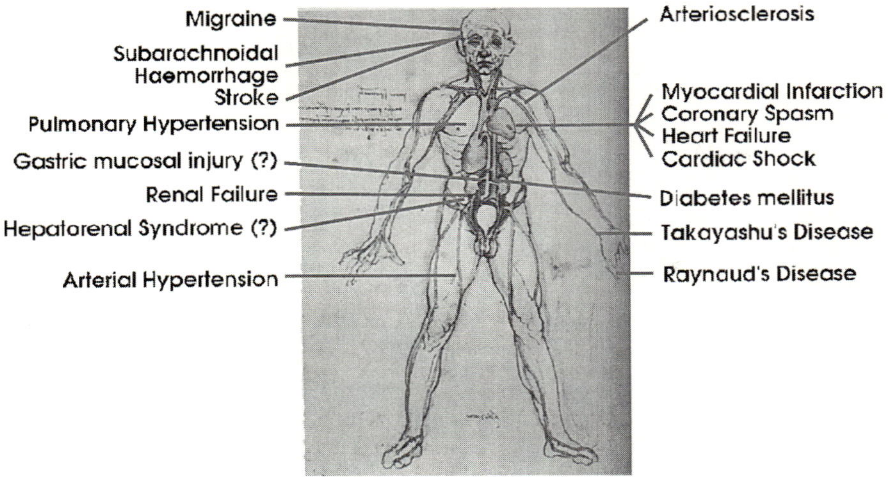

Fig. 2. Endothelin and human disease. Disease states with increased endothelin production are shown, as judged from plasma endothelin levels [14]

Quite sustained increases in plasma endothelin levels can be observed in heart failure [16–21]. Most interestingly, the degree of increase in plasma endothelin levels is associated with pulmonary arterial pressures [22], suggesting that congestion of the lungs occurring in heart failure is a major component in explaining activation of the endothelin vasopressor axis. It remains uncertain, however, whether the increased local vascular pressure, the decreased oxygen content of the blood or activation of other vasopressor systems is the exact underlying cause.

As thrombin and ischemia increase endothelin production and in part also upregulation of its receptors, one would expect activation of this pressure system in acute coronary syndromes. Indeed, in patients with unstable angina, increased staining for endothelin can be demonstrated in coronary lesions removed during atherectomy [23]. Similarly, in patients with acute myocardial infarction, endothelin plasma levels are increased already at the time of hospitalization and then decline over the next few days in the patients of the low-risk group (lower mean age, less anterior infarction, higher left ventricular ejection fraction). In contrast, in the group of patients after acute myocardial infarction with high risk (i.e., higher mean age, reduced ejection fraction, higher incidence of anterior myocardial infarction) plasma endothelin levels remain elevated for prolonged periods of time [24]. It is likely that in the setting of acute myocardial infarction, ischemia (and possibly reperfusion) as well as activation of the coagulation cascade with thrombin formation is responsible for the increased production of the peptide. Indeed, both hypoxia and thrombin are important stimulators of endothelin production.

Although hypertension is an attractive candidate disease for a role of endothelin, the currently available evidence is quite controversial (Table 1; see [25, 26]). Plasma endothelin levels are normal in most forms of hypertension, although they may be increased in certain patients and in particular in those with atherosclerotic vas-

Table 1. Arguments for and against a role of endothelin (ET) in hypertension

For	Against
ET induces persistent hypertension	Plasma ET levels mostly normal in hypertension
ET-secreting tumors induce hypertension	Vascular response to ET reduced in hypertension
Vascular ET increased in DOCA hypertension	Vascular ET normal in SHR
ET plasma levels increased in certain hypertensive patients	ET-2 transgenic rats normotensive
Sensitivity of renal arteries to ET increased in SHR	ET knockout mice hypertensive
ET antagonists lower blood pressure in SHR	

DOCA, desoxycorticosterone acetate; SHR, spontaneously hypertensive rats

cular disease and/or renal failure. In patients with malignant hemangioendotheliomas, marked increases in plasma endothelin levels as well as hypertension have been described (Fig. 3) [27]. Both the increase in plasma endothelin levels and arterial blood pressure are reversible after removal of the tumor in these patients [27]. Hence, this clinical syndrome demonstrates that increased vascular endothelin production in principle can cause hypertension in man. In experimental models of hypertension, endothelin production has been shown to be increased in desoxycorticosterone acetate (DOCA) salt hypertension [28], while it appears normal or even reduced in the spontaneously hypertensive rat. In the aorta as well as mesenteric resistance arteries (but not in the renal circulation) of various forms of experimental hypertension, the response to endothelin is depressed, most likely due to down-regulation of endothelin receptors. Endothelin antagonists, particularly those blocking both ET_A and ET_B receptors are able to lower blood pressure in rats with DOCA salt-induced hypertension, while the results in the spontaneously hypertensive rat are controversial [29]. Studies in patients with essential hypertension are ongoing and they will demonstra-

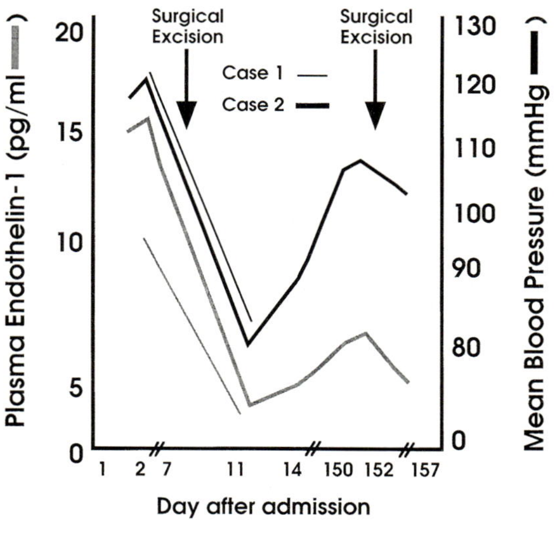

Fig. 3. The production of high amounts of endothelin from endothelial tumors such as hemangioendotheliomas with subsequently markedly elevated plasma levels of the peptide is associated with elevated arterial blood pressure. Both blood pressure and endothelin plasma levels normalize after surgical excision of the tumor, return to high levels upon reappearance of the tumor, and fall again after the second surgical excision. (Modified from [27])

te whether or not vascular endothelin production contributes to the increase in vascular tone occurring in hypertension.

In contrast to arterial hypertension, pulmonary hypertension clearly is associated with marked increases in plasma endothelin levels as well as an increased vascular production of endothelin, also in the human [22, 30]. Indeed, even in healthy subjects, high altitude exposure is associated with doubling of plasma endothelin levels (Fig. 4) [30]. Most interestingly, the increase in plasma endothelin levels in healthy mountaineers at high altitude is significantly correlated with pulmonary systolic pressure, and both changes are reversible upon oxygen breathing. Hence, it is likely that endothelin plays an important role in the regulation of perfusion under conditions of hypoxia in the lung.

Low-tension glaucoma is a disease state of the ophthalmic circulation in which intraocular pressure is normal (in contrast to the classic forms of glaucoma), but similar changes in the optic nerve with their clinical consequences occur [2]. Enhanced vascular tone of the retinal circulation with local tissue ischemia has been accounted for these changes in low-tension glaucoma. Although one report suggests an increase in plasma endothelin levels in these patients, we were unable to confirm those findings. However, it appears that the physiological increase in plasma endothelin levels upon changes in posture from the supine to the upright position is abnormal in patients with low-tension glaucoma [31].

Furthermore, in kidney disease such as acute renal failure, several experimental studies suggest a pathophysiologic role of endothelin. Indeed, acute ischemic renal failure leads to an increase in endothelin release and/or endothelin receptor upregulation [32–35]; this effect can be reversed by endothelin receptor antagonists or drugs blocking the endothelin-converting enzyme [36, 37]. In several other situations with

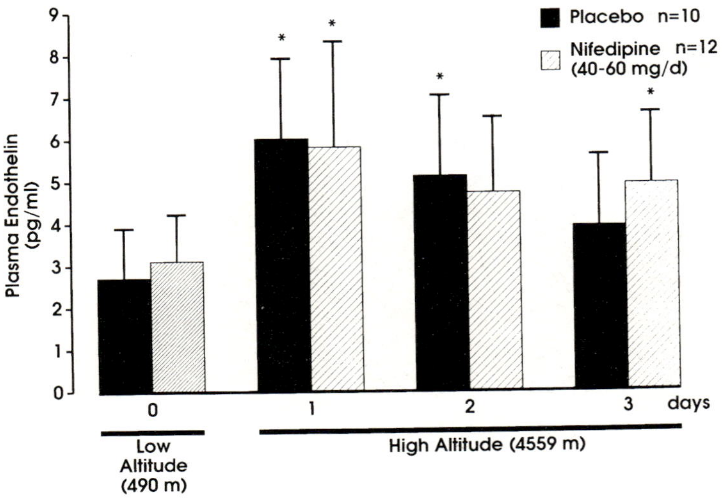

Fig. 4. Plasma levels of endothelin increase in healthy volunteers as they reach high altitude and remain at high altitude over 3 days. The calcium channel blocker nifedipine does not influence endothelin plasma levels under these conditions. (Modified from ([30])

primary or secondary involvement of the kidney, elevated endothelin plasma levels have been described, such as uremia [38], renal failure induced by nephrotoxic agents, i.e. radiocontrast agents [18, 39–42], amphotericin B [39] and possibly cisplatin [43], whereas immunosuppressive agents such as cyclosporin and FK 506 seem to modulate endothelin release [44–48]. Obviously, acute renal failure can be caused by a variety of stimuli and the importance of a given mediator may vary depending on the major cause involved. Ischemic renal failure is relatively well defined experimentally and is particularly suitable for studies with newly developed drugs. Indeed, in a monkey model, decreases in renal blood flow in acute ischemic renal failure could be prevented by an ET_A/ET_B receptor antagonist [93].

Calcium Antagonists

Three main classes of calcium antagonists have been delineated: (1) phenylalkylamine-type (prototype: verapamil) calcium antagonists, (2) dihydropyridine-type (prototype: nifedipine), and (3) benzothiazepine-type (prototype: diltiazem) calcium antagonists (Fig. 5). All substances interfere with different parts of the voltage-operated calcium channels, thereby decreasing influx of extracellular calcium into cells, in particular vascular smooth muscle cells. Increases in intracellular calcium are a very basic biological signal mechanism involved in numerous intracellular processes [5]. In the context of coronary artery disease, increases in intracellular calcium are particularly important in platelets (leading to platelet activation), in vascular smooth muscle cells (leading to vasoconstriction) as well as in endothelial cells (leading to the release of vasoactive substances).

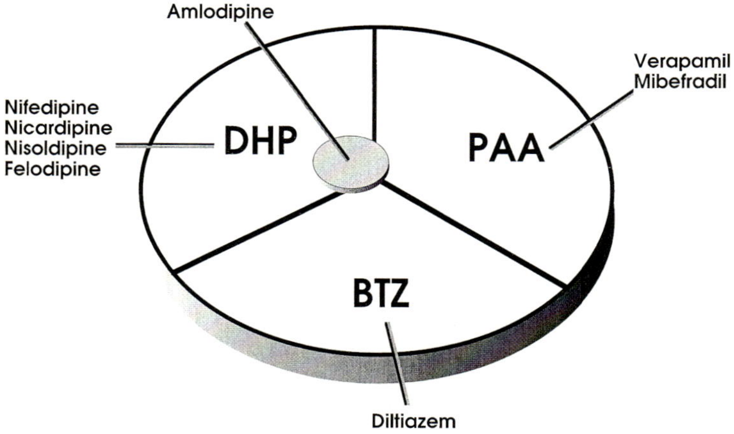

Fig. 5. Types of calcium channel blockers depending of the binding site to the α_1-subunit of the L-type channels. *DHP*, dihydropyridine type; *PAA*, phenylalkylamine type; *BTZ*, benzothiazepine type. (Modified from [92])

Calcium Antagonists and Endothelium-Dependent Relaxations

The formation of nitric oxide from L-arginine via the enzyme nitric oxide synthase [50] is associated with increases in intracellular calcium [51–53]. Indeed, the calcium ionophore A23187 is a very potent activator of the release of this relaxing factor under most conditions. Increases in intracellular calcium are most likely important for the release of nitric oxide (Fig. 1) [51]; under most conditions, however, calcium antagonists do not affect the release of this mediator as endothelial cells do not appear to have voltage-operated calcium channels [54].

However, calcium antagonists facilitate the effects of nitric oxide at the level of vascular smooth muscle. Indeed, nitric oxide exerts its effects via guanylyl cyclase leading to increases in intracellular cyclic guanosine monophosphate (cGMP) concentrations [55]. cGMP lowers intracellular calcium (most likely by increasing calcium efflux and reuptake of calcium into intracellular stores) and dephosphorylates myosin light chains. In vessels treated with calcium antagonists, endothelium-dependent relaxations and those to the endothelium-independent vasodilator sodium nitroprusside (which, like nitric oxide, also causes relaxation via cGMP) are augmented in line with the previous interpretation of the effects of calcium antagonists [54, 56].

Calcium Antagonists and Endothelin

The production of endothelin in endothelial cells is associated with an increase in intracellular calcium. The calcium ionophore A 23187, which increases intracellular calcium levels in endothelial cells, is a very potent stimulator of endothelin production [57]. However, calcium antagonists do not alter endothelin production in vitro nor in vivo. Indeed, in healthy mountaineers, plasma endothelin levels increased substantially at high altitude; this increase is not modulated by administration of a high dose of nifedipine (Fig. 4) [30].

Although originally endothelin was considered an endogenous activator of voltage-operated calcium channels [4], more recent data have demonstrated that endothelin interacts with specific receptors on vascular smooth muscle (i.e. ET_A and ET_B receptors) mediating vasoconstriction and proliferation as well as on endothelial cells (i.e, ET_B receptors linked to the formation of nitric oxide and prostacyclin; Fig. 1) [59, 60]. In certain blood vessels such as the porcine coronary artery the endothelin receptors on vascular smooth muscle are linked to voltage-operated calcium channels via a G_i protein [61]. This may explain why calcium antagonists do reduce endothelin-induced vasoconstriction in this blood vessel. In other blood vessels such as the human internal mammary artery on the other hand [62], most of the contractile effects induced by endothelin are mediated by release of intracellular calcium (after activation of phospholipase C with formation of inositol trisphosphate and diacylglycerol) [63, 64]. This may explain why calcium antagonists of all classes are unable to prevent endothelin-induced contractions in this artery [62]. On the other hand, in most blood vessels, calcium antagonists are able to reverse endothelin-induced contractions [62]. This most likely is related to the fact that endothelin lowers membrane potential of vascular smooth muscle cells [65] and thereby opens voltage-operated calcium channels. Hence, once contractions have developed, calcium antagonist do exert an inhibitory effect due to this phenomenon.

An interesting other aspect of endothelin is its capacity to potentiate contractions of other agonists such as serotonin and norepinephrine even at subthreshold concentrations [66, 67]. Indeed, in the human internal mammary and coronary artery, serotonin and norepinephrine (in the case of the mammary artery) markedly augment the contractile responses to these agonists [67]. These indirect potentiating effects of endothelin are due to an increased calcium sensitivity of vascular smooth muscle (Fig. 6) and are therefore prevented by treatment with calcium antagonists of the dihydropyridine type (Fig. 7) [67].

Several studies have suggested that the small vessels are more dependent on the influx effects of cellular calcium than larger vessels [68]. This may explain why in the human forearm circulation intraarterial application of verapamil or nifedipine are able to prevent endothelin-induced contractions (Fig. 8) [11]. Indeed, under these conditions the vasodilator effects of endothelin are most prominent. However, particularly with higher concentrations of endothelin-1, oral administration of verapamil does not appear to be effective enough to reduce endothelin-induced contractions (W. Kiowski, L. Linder, T. F. Lüscher 1992, unpublished observations). Hence, higher concentrations of the drugs may be required to fully inhibit vasoconstriction induced by this potent biological peptide. It is uncertain at this point whether clinically used dosages of verapamil or nifedipine are effectively inhibiting endogenously formed endothelin.

We developed a double-injection model which allows to investigate the effects of vasoactive substances in the human skin microcirculation in vivo using laser Doppler flowmetry (Fig. 9) [69]. This experimental set-up allows to determine relative changes in skin blood flow after injection of endothelin-1 in the presence or absence

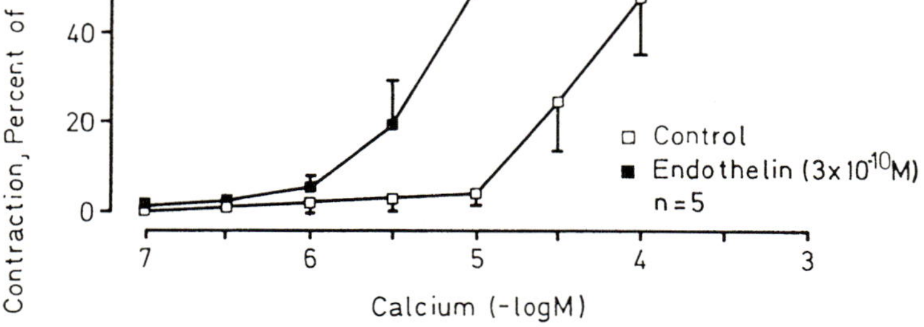

Fig. 6. In human mammary arteries studied in vitro, calcium contraction curves were drawn after removal of intracellular calcium (by caffeine) and extracellular calcium (by calcium-free solutions). Endothelin at subthreshold concentrations enhances the vasoconstriction response to calcium. (Modified from [67], with permission)

of calcium channel blockers or endothelin receptor antagonists. Systemic administration of diltiazem (240 mg p.o.) slightly tended to reduce the vasoconstriction of endothelin-1 (Fig. 10).

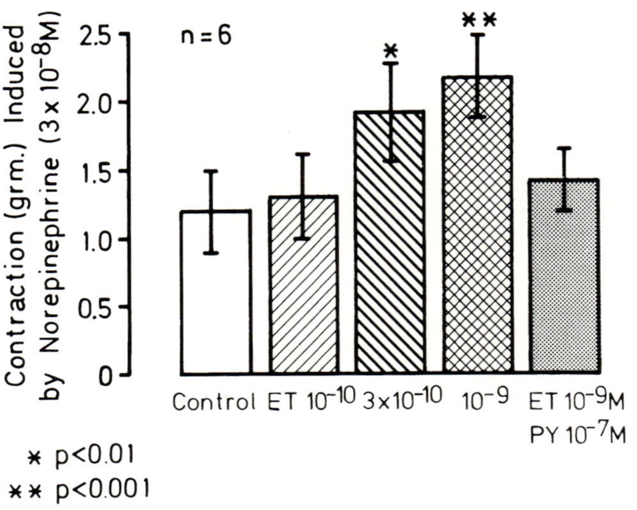

Fig. 7. The contraction response of human mammary arteries to norepinephrine can be enhanced by subthreshold concentrations of endothelin (*ET*). The dihydropyridine calcium antagonist darodipine (*PY*) prevents this effect, without affecting the contraction to norepinephrine. (Modified from [67], with permission)

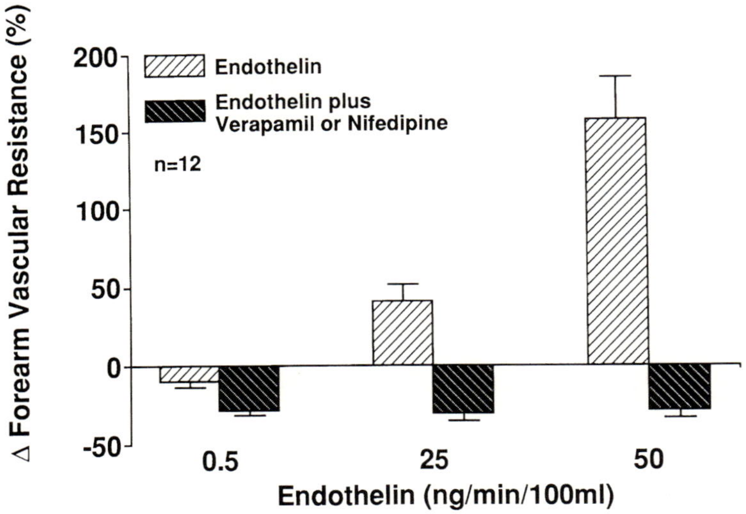

Fig. 8. Intraarterial infusion of endothelin-1 into the human forearm circulation at low doses leads to a slight vasodilation; however, higher doses induce marked and long-lasting vasoconstriction. Intraarterial administration of calcium channel blockers such as verapamil or nifedipine can fully prevent the vasoconstriction response to endothelin. (Modified from [11], with permission)

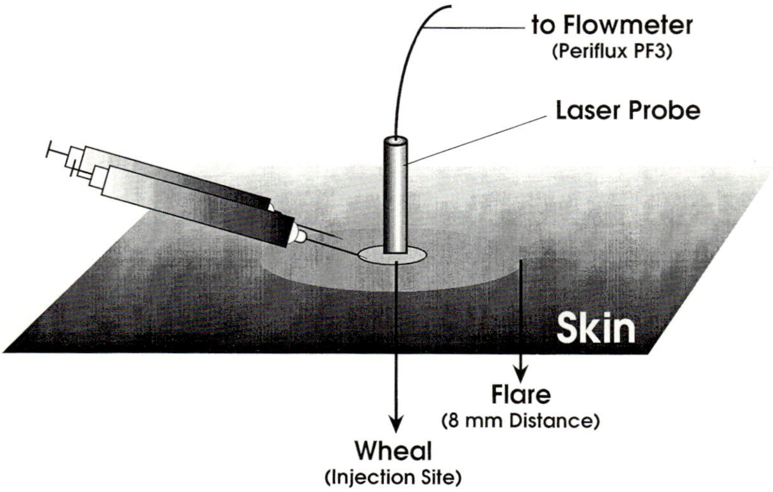

Fig. 9. Double-injection model for the investigation of changes in blood flow in the human skin microcirculation in vivo using laser Doppler flowmetry. The model allows the intradermal simultaneous injection of two drugs, i.e., antagonist and agonist, thus allowing the effects and interactions of drugs and newly synthesized molecules in very small amounts to be assessed without systemic effects. Changes in skin blood flow are measured at the injection site and the surrounding area (wheal)

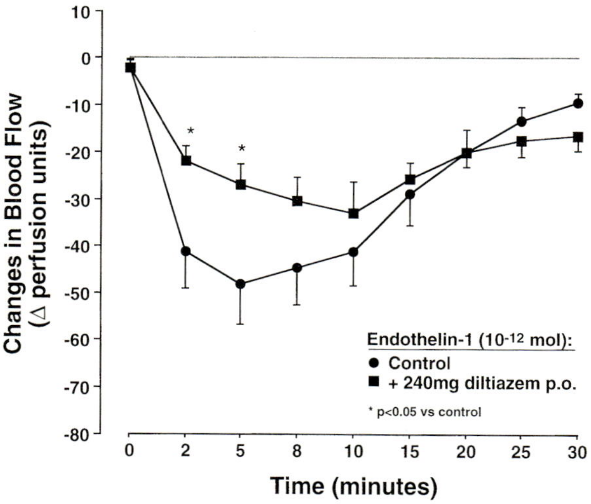

Fig. 10. Effects of the calcium channel blocker diltiazem (240 mg p.o.) 4 h after drug intake on exogenous endothelin in the human skin microcirculation in vivo in patients with coronary artery disease, as assessed by laser Doppler flowmetry. Endothelin leads to a marked vasoconstriction. The oral application of the calcium channel blocker diltiazem slightly inhibits endothelin-induced vasoconstriction

Calcium Antagonists and Vascular Smooth Muscle

In damaged blood vessels or those with functionally altered endothelial cells, platelets adhere quite rapidly. Platelets are an important source of platelet-derived growth factor, a potent mitogen in a variety of cells [70]. In human coronary arteries, smooth muscle cells held in culture, platelet-derived growth factor induces marked proliferative responses with a half-maximal concentration of 0.1–0.5 ng/ml [71]. The effect of platelet-derived growth factor involves activation of tyrosine kinase intracellularly [72]. Of particular interest is the fact that verapamil in a concentration which may be achieved in patients treated with the drug inhibits platelet-derived growth factor induced proliferative responses in human coronary vascular smooth muscle cells [71]. In contrast, proliferative responses induced by pulsatile stretch are unaffected by either verapamil or nifedipine. This indicates that mechanical forces such as pulsatile stretch and receptor-operated growth factors such as platelet-derived growth factor operate via distinct mechanisms. The fact that neither the activity of tyrosine kinase nor activation of S6c kinase nor MAP (mitogen-activated protein kinase) is affected by calcium antagonists suggests that they interact at more distal intracellular signal transduction mechanisms [71].

In vivo in cholesterol-fed rabbits several classes of calcium antagonists inhibit the atherosclerotic process [73]. Three large angiographic trials using nifedipine or nicardipine have demonstrated that chronic therapy with a calcium antagonist reduces the development of new lesions in the coronary circulation without affecting already existing stenosis [74]. These clinical studies have confirmed that calcium antagonists do have an antiatherosclerotic action, although it is less pronounced as compared to the angiographic results of lipid lowering therapy [74–76].

Endothelin Receptor Antagonists

With the cloning of endothelin receptors, definition of the vascular actions of endothelin has become well characterized. Several molecules have now been tested on

Table 2. Endothelin (ET) antagonists

Drug	Receptor
BE-18 257 A/B	ET_A
BQ-162	ET_A
BQ-123	ET_A
BQ-153	ET_A
PD 147 953 (=FR 139 317)	ET_A
TAK-044	ET_A/(ET_B)
LU-127 043	ET_A
RO-462 005	ET_A/ET_B
RO-470 203 (bosentan)	ET_A/ET_B
SB 209 670	ET_A/ET_B
PD 142 893	ET_A/ET_B
PD 145 065	ET_A/ET_B

their ability to infere with the vasoconstriction effects of endothelin and more recently endothelin antagonists have been developed (Table 2) [29, 77–88]. While the older molecules are specific and interfere primarily with ET_A receptors, more recently developed molecules can interact with both ET_A and ET_B receptors. Certain of the endothelin antagonists also are orally active and hence suitable for further, now ongoing clinical trials.

Endothelin Antagonists in Isolated Human Blood Vessels

In the isolated human internal mammary artery, endothelin causes profound vasoconstriction. The ET_A receptor antagonist FR 139 317 shifts the concentration–response curve of endothelin-1 to the right. The ET_B receptor mediated part of the vasoconstriction, as obtained with the ET_B agonist sarafotoxin, can only be prevented by the combined ET_A/ET_B antagonist, i.e. bosentan. Indeed, these experiments suggest that ET_B receptors also contribute to endothelin-induced vasoconstriction in human vascular tissue (Fig. 11). Northern blot analysis using cDNA probe for endothelin ET_A receptors further demonstrated that indeed the receptors are expressed in human blood vessels [10]. Most interestingly, pharmacological data strongly suggest, that the contribution of ET_B receptors is more important in veins as compared to arteries [62]. More recently, infusion studies in human hand veins with sarafotoxin S6C confirmed that the ET_B receptor agonist is indeed a potent vasoconstrictor in the human venous circulation [90].

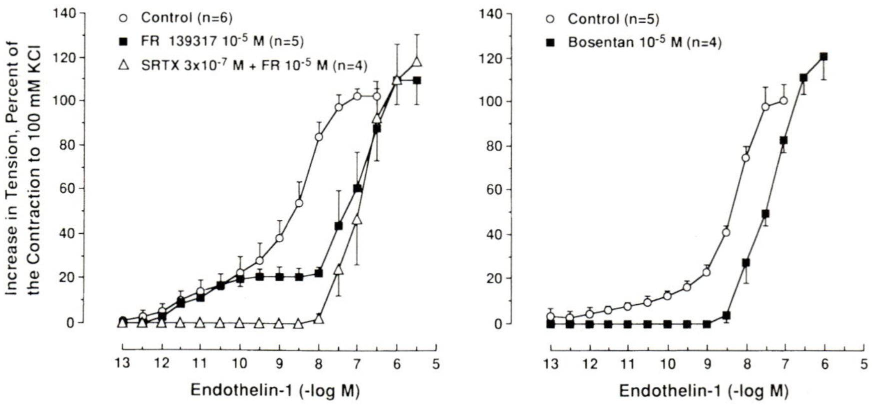

Fig. 11. Contraction responses to endothelin (ET-1) in human mammary resistance arteries. The effects of the ET_A receptor antagonist FR 139 317 (*FR*) alone or in combination with sarafotoxin S6c (*SRTX, left panel*) as well as the effects of the combined ET_A/ET_B antagonist bosentan (*right panel*) are shown. The ET_A antagonist prevents endothelin-induced vasoconstriction only in part. The ET_B receptor-mediated part of the vasocontriction represented by the sarafotoxin-induced vasoconstriction can only be inhibited by the combined ET_A/ET_B antagonist bosentan. (From [89], with permission)

Endothelin Antagonists in Humans In vivo

The human skin microcirculation can be used as a surrogate circulation to test pharmacological agents relatively noninvasively (Fig. 9) [69]. Particularly with new compounds, the risk for subjects is very low and vascular efficacy of molecules can easily be demonstrated without systemic effects. We have used this model to test the efficacy of ET_A and ET_A/ ET_B receptor antagonists in healthy human subjects. Injection of PD 147 953 (a molecule corresponding to FR 139 317) as well as of the ET_A/ET_B receptor antagonist PD 145 065 led to a comparable increase in skin blood flow (Fig. 12, top panel) [69]. The increases in blood flow induced by the injection of the endothelin antagonists alone strongly suggests that endothelin contributes to the regulation of vascular tone in normal men. This conclusion is further supported from the more recent observation that BQ-123, a selective ET_A receptor antagonist, increases forearm blood flow as assessed by plethysmography in normal subjects [91].

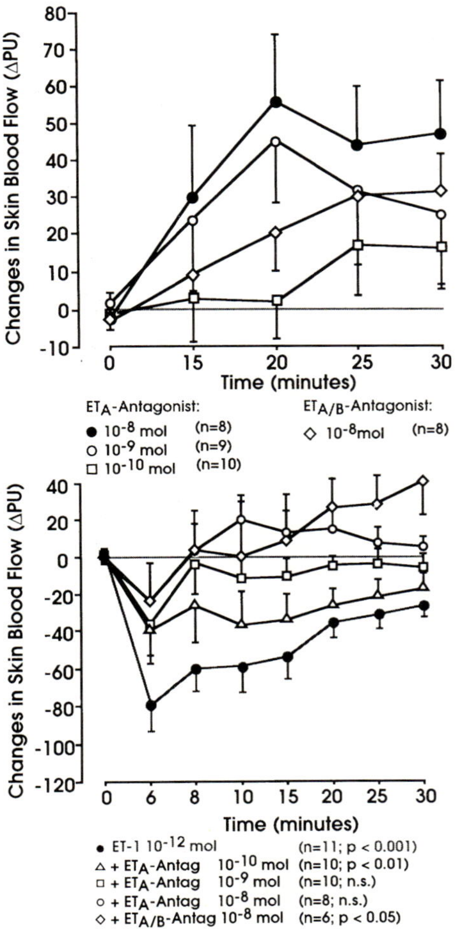

Fig. 12. Effects of the endothelin $(ET)_A$ receptor antagonist PD 147 953 (equivalent to the compound FR 139 317) and the ET_A/ET_B receptor antagonist PD 145 065 on blood flow (*top panel*) and on endothelin-induced vasoconstriction (*lower panel*) in the human skin microcirculation, as assessed by laser Doppler flowmetry. The antagonists increase skin blood flow on their own and fully inhibit endothelin-1-induced vasoconstriction in skin blood flow. *PU*, perfusion units. (Modified from [69])

In the human skin microcirculation, Injection of endothelin-1 alone causes a profound vasoconstriction at the injection site as well as a marked vasodilation around the injection wheel. The endothelin-1-induced vasoconstriction of the human skin microcirculation can be concentration-dependently inhibited by endothelin receptor antagonists (Fig. 12, lower panel) [69]. Similarly, BQ-123 reduces the vasoconstrictor effects of endothelin-1 in the human forearm circulation [91].

Thus, endothelin antagonists in vitro as well as experimentally and in the few published in vivo experiments in humans appear to be potent inhibitors of endothelin induced vasomotion. Ongoing clinical trials will demonstrate their efficacy in the therapy of cardiovascular diseases such as acute myocardial infarction, unstable angina, congestive heart failure, but also renal failure and other disease states with elevated plasma and/or local endothelin levels.

Summary

Abnormal vasoconstriction is a common feature of several diseases of the cardiovascular system, in particular atherosclerosis, hypertension, Raynaud's disease, low-tension glaucoma as well as coronary and cerebral vasospasm. Abnormal neuronal control of vascular tone and circulating hormones have been implicated in these disease states, but proof is still lacking. Endothelin is a new 21-amino acid peptide released by the blood vessel wall which is a potent vasoconstrictor. In the intact blood vessel wall, endothelin-1 is exclusively released by endothelial cells, primarily to the abluminal site. In vascular smooth muscle, endothelin causes vasoconstriction after activation of ET_A- and in part also ET_B-receptors. These endothelin receptors are G-protein linked and associated with activation of phospholipase C with concomitant formation of inositol trisphosphate and diacylglycerol. In addition, activation of endothelin receptors in vascular smooth muscle may open voltage-operated calcium channels and in turn cause profound and sustained vasoconstriction. Endothelin production is upregulated in several diseases of the cardiovascular system and, hence, pharmacological blockade of this vasopressor axis is of pharmacological and clinical interest. Calcium channel blockers inhibit the influx of extracellular calcium through voltage-operated calcium channels into vascular smooth muscle cells. Depending on whether intracellular calcium release or entry of extracellular calcium contributes more to endothelin-induced contractions, calcium channel blockers are more or less efficient in inhibiting endothelin-induced contractions. While in the mammary artery, the drugs are not very effective, they do cause profound inhibition of endothelin-induced vasoconstriction in human coronary arteries, the human forearm circulation, and human skin microcirculation. Endothelin receptor antagonists occupy either ET_A and/or ET_A-/ET_B receptors. Several compounds have been synthesized; the drugs are extremely effective in inhibiting endothelin-induced vasoconstriction in vitro and in isolated perfused organs in vivo as well. In the human, they have only been tested in the skin and forearm circulation. A direct comparison in the human skin microcirculation revealed a comparable efficacy of calcium channel blockers and endothelin antagonists, although the latter were somewhat more effective.

In conclusion, calcium channel blockers and endothelin antagonists are most effective in interfering with endothelin-incluced vasoconstriction in man. Ongoing clinical trials will determine the importance of such pharmacological properties to treat cardiovascular disease in patients.

Acknowledgement. Research was supported by grants the Swiss National Research Foundation (No. 32-32562.91), the German Research Association (Deutsche Forschungsgemeinschaft DFG, WE 1772/1-1) and a grant-in-aid from Warner Lambert Gödecke Parke Davis, Freiburg, Germany.

References

1. Lüscher TF, Vanhoutte PM (1990) The endothelium: modulator of cardiovascular function. CRC Press, Boca Raton
2. Gasser P, Flammer J (1987) Influence of vasospasm on visual function. Doc Ophthalmol 66:3–18
3. Kassell NF, Sasaki T, Colohan ART, Nazar GB (1985) Cerebral vasospasm following aneurysmal subarachnoid hemorrhage. Stroke 16:562–572
4. Yanagisawa M, Kurihara H, Kimura S, Tomobe Y, Kobayashi M, Mitsui Y, Yazaki Y, Goto K, Masaki T (1988) A novel potent vasoconstrictor peptide produced by vascular endothelial cells. Nature 332:411–415
5. Yanagisawa M, Masaki T (1989) Molecular biology and biochemistry of the endothelins. Trends Pharmacol Sci 10:374–378
6. Yanagisawa M, Inoue A, Ishikawa T, Kasuya Y, Kimura S, Kumagaye S, Nakajima K, Watanabe TX, Sakakibara S, Goto K, Masaki T (1988) Primary structure, synthesis, and biological activity of rat endothelin, an endothelium-derived vasoconstrictor peptide. Proc Natl Acad Sci USA 85: 6964–6967
7. Wagner OF, Christ G, Wojta J, Vierhapper H, Parzer S, Nowotny PJ, Schneider B, Waldhäusl W, Binder BR (1992) Polar secretion of endothelin-1 by cultured endothelial cells. J Biol Chem 267:16066–16068
8. Kurihara T, Akimoto M, Kurokawa K, Ishiguro H, Niimi A, Maeda A, Sigemoto M, Yamashita K, Yokoyama I, Hirayama Y et al (1992) ET-3 sensitive reduction of tissue blood flow in rat liver. Life Sci 51:6
9. Karne S, Jayawickreme CK, Lerner MR (1993) Cioning and characterization of an endothelin-3 specific receptor (ETC receptor) from Xenopus laevis dermal melanophores. J Biol Chem 268:19126–19133
10. Seo B, Oemar BS, Siebenmann R, von Segesser L, Lüscher TF (1994) Both ETA and ETB receptors mediate contraction to endothelin-1 in human blood vessels. Circulation 89:1203–1208
11. Kiowski W, Lüscher TF, Linder L, Buhler FR (1991) Endothelin-1-induced vasoconstriction in humans. Reversal by calcium channel blockade but not by nitrovasodilators or endothelium-derived relaxing factor. Circulation 83:469–475
12. Lüscher TF, Yang Z, Tschudi M, Von SL, Stulz P, Boulanger C, Siebenmann R, Turina M, Buhler FR (1990) Interaction between endothelin-1 and endothelium-derived relaxing factor in human arteries and veins. Circ Res 66:1088–1094
13. Lüscher TF, Boulanger CM, Dohi Y, Yang ZH (1992) Endothelium-derived contracting factors. Hypertension 19:117–130
14. Lüscher TF (1993) Do we need endothelin antagonists? Cardiovasc Res 27:2089–2093
15. Zamora MR, O'Brien RF, Rutherford RB, Weil JV (1990) Serum endothelin-1 concentrations and cold provocation in primary Raynaud's phenomenon (see comments). Lancet 336:1144–1147
16. Cavero PG, Miller WL, Heublein DM, Margulies KB, Burnett JJ (1990) Endothelin in experimental congestive heart failure in the anesthetized dog. Am J Physiol (Renal Fluid Electrolyte Physiol) 259:28–32

17. Cernacek P, Stewart DJ (1989) Immunoreactive endothelin in human plasma: marked elevations in patients in cardiogenic shock. Biochem Biophys Res Commun 161:562–567
18. Margulies KB, Hildebrand FJ, Lerman A, Perrella MA, Burnett JJ (1990) Increased endothelin in experimental heart failure. Circulation 82:2226–2230
19. Noll G, Tschudi MR, Novosel D, Lüscher TF (1994) Activity of the L-arginine/nitric oxide pathway and endothelin-1 in experimental heart failure. J Cardiovasc Pharm 23:916–921
20. Rodeheffer RJ, Lerman A, Heublein DM, Burnett JJ (1992) Increased plasma concentrations of endothelin in congestive heart failure in humans (see comments). Mayo Clin Proc 67:719–724
21. Stewart DJ, Cernacek P, Costello KB, Rouleau JL (1992) Elevated endothelin-1 in heart failure and loss of normal response to postural change. Circulation 85:510–517
22. Cody RJ, Haas GJ, Binkley PF, Capers Q, Kelley R (1992) Plasma endothelin correlates with the extent of pulmonary hypertension in patients with chronic congestive heart failure. Circulation 85:504–509
23. Zeiher A (1994) Endothelin in patients with unstable angina. Lancet 334:1405–1406
24. Stewart DJ, Kubac G, Costello KB, Cernacek P (1991) Increased plasma endothelin-1 in the early hours of acute myocardial infarction. J Am Coll Cardiol 18:38–43
25. Lüscher TF, Seo BG, Bühler FR (1993) Potential role of endothelin in hypertension: Controversy of endothelin in hypertension. Hypertension 21:752–757
26. Vanhoutte PM (1993) Is endothelin involved in the pathogenesis of hypertension? Hypertension 21:747–751
27. Yokokawa K, Tahara H, Kohno M, Murakawa K, Yasunari K, Nakagawa K, Hamada T, Otani S, Yanagisawa M, Takeda T (1991) Hypertension associated with endothelin-secreting malignant hemangioendothelioma. Ann Intern Med 114:213–215
28. Larivière R, Thibault G, Schiffrin EL (1993) Increased endothelin-1 content in blood vessels of deoxycorticosterone acetate-salt hypertensive rats but not in spontaneously hypertensive rats. Hypertension 21:294–300
29. Nishikibe M, Tsuchida S, Okada M, Fukuroda T, Shimamoto K, Yano M, Ishikawa K, lkemoto F (1993) Antihypertensive effect of a newly synthesized endothelin antagonist, BQ-123, in a genetic hypertensive model. Life Sci 52:717–724
30. Goerre S, Wenk M, Bärtsch P, Lüscher TF, Niroomand S, Hohernhaus H, Hoelz O, Reinhart WH (1995) Endothelin-1 in pulmonary hypertension associated with high altitude. Circulation 90: 359–364
31. Kaiser H, Flammer J, Wenk M, Lüscher TFI (1995) Abnormal response of endothelin-1 plasma levels to changes in posture in patients with low tension glaucoma. Arch Ophtalmol (in press)
32. Roubert P, Cornet S, Plas P, Guilmard C, Pirotzky E, Chabrier PE, Braquet P (1993) Upregulation of renal endothelin receptors in glycerol-induced acute renal failure in the rat. J Cardiovasc Pharmacol 22:S303–S305
33. Tomita K, Ujiie K, Nakanishi T, Tomura S, Matsuda O, Ando K, Shichiri M, Hirata Y, Marumo F (1989) Plasma endothelin levels in patients with acute renal failure (letter). N Engl J Med 321:16
34. Warrens AN, Cassidy MJ, Takahashi K, Ghatei MA, Bloom SR (1990) Endothelin in renal failure. Nephrol Dial Transplant 5:418–422
35. Nambi P, Pullen M, Contino LC, Brooks DP (1990) Upregulation of renal endothelin receptors in rats with cyclosporine A-induced nephrotoxicity. Eur J Pharmacol 187:113–116
36. Gellai M, Jugus M, Fletcher T, De Wolf R, Nambi P (1994) Reversal of postischemic acute renal failure with a selective endothelin. A receptor antagonist in the rat. J Clin Invest 93:900–906
37. Vemulapalli S, Chiu PJ, Chintala M, Bernardino V (1993) Attenuation of ischemic acute renal failure by phosphoramidon in rats. Pharmacology 47:188–193
38. Mallamaci F, Parlongo S, Zoccali C (1993) Influence of cardiovascular damage and residual renal function on plasma endothelin in chronic renal failure. Nephron 63:291–295
39. Heyman SN, Clark BA, Kaiser N, Epstein FH, Spokes K, Rosen S, Brezis M (1992) In-vivo and in-vitro studies on the effect of amphotericin B on endothelin release. J Antimicrob Chemother 29:69–77
40. Heyman SN, Clark BA, Kaiser N, Spokes K, Rosen S, Brezis M, Epstein FH (1992) Radiocontrast agents induce endothelin release in vivo and in vitro. J Am Soc Nephrol 3:58–65
41. Heyman SN, Clark BA, Cantley L, Spokes K, Rosen S, Brezis M, Epstein FH (1993) Effects of ioversol versus iothalamate on endothelin release and radiocontrast nephropathy. Invest Radiol 28:313–318

42. Cantley LG, Spokes K, Clark B, McMahon EG, Carter J, Epstein FH (1993) Role of endothelin and prostaglandins in radiocontrast-induced renal artery constriction. Kidney Int 44:1217–1223
43. Ohta K, Hirata Y, Shichiri M, Ichioka M, Kubota T, Marumo F (1991) Cisplatin-induced urinary endothelin excretion (letter). JAMA 265:1391–1392
44. Ong AC, Jowett TP, Scoble JE, O'Shea JA, Varghese Z, Moorhead JF (1993) Effect of cyclosporin A on endothelin synthesis by cultured human renal cortical epithelial cells. Nephrol Dial Transplant 8:748–753
45. Bunchman TE, Brookshire CA (1991) Cyclosporine-induced synthesis of endothelin by cultured human endothelial cells. J Clin Invest 88:310–314
46. Iwasaki S, Homma T, Kon V (1994) Site specific regulation in the kidney of endothelin and its receptor subtypes by cyclosporine. Kidney Int 45:592–597
47. Copeland KR, Yatscoff RW (1992) Comparison of the effects of cyclosporine and its metabolites on the release of prostacyclin and endothelin from mesangial cells. Transplantation 53: 640-645
48. Moutabarrik A, Ishibashi M, Fukunaga M, Kameoka H, Kawaguchi N, Takano Y, Kokado Y, Sonoda T, Onishi S, Takahara S et al (1992) FK506-induced kidney tubular cell injury. Transplantation 54:1041–1047
49. Tschudi M, Richard V, Bühler FR, Lüscher TF (1991) Importance of endothelium-derived nitric oxide in porcine coronary resistance arteries. Am J Physiol 260:H13–H20
50. Furchgott RF, Zawadzki JV (1980) The obligatory role of endothelial cells in the relaxation of arterial smooth muscle by acetylcholine. Nature 288:373–376
51. Palmer RM, Ferrige AG, Moncada S (1987) Nitric oxide release accounts for the biological activity of endothelium-derived relaxing factor. Nature 327:524–526
52. Busse R, Fichtner H, Luckhoff A, Kohlhardt M (1988) Hyperpolarization and increased free calcium in acetylcholine-stimulated endothelial cells. Am J Physiol 255:H965–H969
53. Palmer RM, Ashton DS, Moncada S (1988) Vascular endothelial cells synthesize nitric oxide from L-arginine. Nature 333:664–666
54. Yang ZH, Von SL, Bauer E, Stulz P, Turina M, Lüscher TF (1991) Different activation of the endothelial L-arginine and cyclooxygenase pathway in the human internal mammary artery and saphenous vein. Circ Res 68:52–60
55. Lüscher TF, Diederich D, Siebenmann R, Lehmann K, Stulz P, Von SL, Yang ZH, Turina M, Gradel E, Weber E et al. (1988) Difference between endothelium-dependent relaxation in arterial and in venous coronary bypass grafts. N Engl J Med 319:462–467
56. Richard V, Tanner FC, Tschudi M, Lüscher TF (1990) Different activation of L-arginine pathway by bradykinin, serotonin, and clonidine in coronary arteries. Am J Physiol 260:H13–H20
57. Boulanger C, Lüscher TF (1990) Release of endothelin from the porcine aorta. Inhibition of endothelium-derived nitric oxide. J Clin 85:587–590
58. Arai H, Hori S, Aramori I, Ohkubo H, Nakanishi S (1990) Cloning and expression of a cDNA encoding an endothelin receptor. Nature 348:730–732
59. Sakurai T, Yanagisawa M, Takuwa Y, Miyazaki H, Kimura S, Goto K, Masaki T (1990) Cloning of a cDNA encoding a non-isopeptide-selective subtype of the endothelin receptor. Nature 348: 732–735
60. Vane J (1990) Endothelins come home to roost (news; comment). Nature 348:973–975
61. Goto K, Kasuya Y, Matsuki N, Takuwa Y, Kurihara H, Ishikawa T, Kimura S, Yanagisawa M, Masaki T (1989) Endothelin activates the dihydropyridine-sensitive, voltage-dependent Ca(2+) channel in vascular smooth muscle. Proc Natl Acad Sci U S A 86:3915–3918
62. Yang Z, Bauer E, Von SL, Stulz P, Turina M, Lüscher TF (1990) Different mobilization of calcium in endothelin-1-induced contractions in human arteries and veins: effects of calcium antagonists. J Cardiovasc Pharmacol 16:654–660
63. Resink TJ, Scott-Burden T, Bühler FR (1988) Endothelin stimulates phospholipase C in cultured vascular smooth muscle cells. Biochem Biophys Res Commun 157:1360–1368
64. Wallnöfer A, Weir S, Ruegg U, Cauvin C (1989) The mechanism of action of endothelin-1 as compared with other agonists in vascular smooth muscle. J Cardiovasc Pharmacol 13 [Suppl 5]: S23–S31
65. Miller VM, Komori K, Burnett JC Jr, Vanhoutte PM (1989) Differential sensitivity to endothelin in canine arteries and veins. Am J Physiol 257:H1127–H1131

66. Tabuchi Y, Nakamaru M, Rakugi H, Nagano M, Ogihara T (1989) Endothelin enhances adrenergic vasoconstriction in perfused rat mesenteric arteries. Biochem Biophys Res Commun 159: 1304–1308
67. Yang Z, Richard V, Von SL, Bauer E, Stulz P, Turina M, Lüscher TF (1990) Threshold concentrations of endothelin-1 potentiate contractions to norepinephrine and serotonin in human arteries. A new mechanism of vasospasm? Circulation 82:188–195
68. Cauvin C, Tejerina M, Hwang O, Kai-Yamamoto M, van Breemen C (1988) The effects of Ca2+-antagonists on isolated rat and rabbit mesenteric resistance vessels. What determines the sensitivity of agonist-activated vessels to Ca2+-antagonists? Ann N Y Acad Sci S22:338–350
69. Wenzel RR, Noll G, Lüscher TF (1994) Endothelin receptor antagonists inhibit endothelin in human skin microcirculation. Hypertension 23:581–586
70. Garg UC, Hassid A (1989) Nitric oxide-generating vasodilators and 8-bromo-cyclic guanosine monophosphate inhibit mitogenesis and proliferation of cultured rat vascular smooth muscle cells. J Clin Invest 83:1774–1777
71. Yang Z, Noll G, Lüscher TF (1993) Calcium antagonists differently inhibit proliferation of human coronary smooth muscle cells in response to pulsatile stretch and platelet-derived growth factor. Circulation 88:832–836
72. Ullrich A, Schlessinger J (1990) Signal transduction by receptors with tyrosine kinase activity. Cell 61:203–212
73. Henry PD, Bentley KI (1981) Suppression of atherogenesis in cholesterol-fed rabbit treated with nifedipine. J Clin Invest 68:1366–1369
74. Lichtlen PR, Hugenholtz P, Rafflenbeul W, Jost S, Deckers JW (1990) Retardation of angiographic progression of coronary artery disease by nifedipine. Lancet 335:1109–1113
75. Blankenhorn DH, Azen SP, Kramsch DM, Mack WJ, Cashin-Hemphill L, Hodis HN, De Boer LW, Mahrer PR, Masteller MJ, Vailas LI (1993) Coronary angiographic changes with lovastatin therapy. The Monitored Atherosclerosis Regression Study (MARS). The MARS Research Group. Ann Intern Med 119:969–976
76. Dumont JM (1993) Effect of cholesterol reduction by simvastatin on progression of coronary atherosclerosis: design, baseline characteristics, and progress of the Multicenter Anti-Atheroma Study (MAAS). Control Clin Trials 14:209–228
77. Aramori I, Nirei H, Shoubo M, Sogabe K, Nakamura K, Kojo H, Notsu Y, Ono T, Nakanishi S (1993) Subtype selectivity of a novel endothelin antagonist, FR 139 317, for the two endothelin receptors in transfected Chinese hamster ovary cells. Mol Pharmacol 43:127–131
78. Breu V, Löffler BM, Clozel M (1993) In vitro characterization of Ro 46-2005, a novel synthetic non-peptide endothelin antagonist of ETA and ETB receptors. FEBS Lett 334:210–214
79. Fujimoto M, Mihara S, Nakajima S, Ueda M, Nakamura M, Sakurai K (1992) A novel non-peptide endothelin antagonist isolated from bayberry, Myrica cerifera. FEBS Lett 305:41–44
80. Fukuroda T, Nishikibe M, Ohta Y, Ihara M, Yano M, Ishikawa K, Fukami T, Ikemoto F (1992) Analysis of responses to endothelins in isolated porcine blood vessels by using a novel endothelin antagonist, BQ-153. Life Sci 50:12
81. Fukuroda T, Nishikibe M, Ihara M, Yano M, Ishikawa K, Ikemoto F (1992) Effects of a novel endothelin antagonist, BO-153, on endothelin-induced vasoconstriction in porcine blood vessels. Jpn J Pharmacol 58:232–240
82. Ihara M, Noguchi K, Saeki T, Fukuroda T, Tsuchida S, Kimura S, Fukami T, Ishikawa K, Nishikibe M, Yano M (1992) Biological profiles of highly potent novel endothelin antagonists selective for the ETA receptor Life Sci 50:247–255
83. Ishikawa K, Fukami T, Nagase T, Fujita K, Hayama T, Niiyama K, Mase T, Ihara M, Yano M (1992) Cyclic pentapeptide endothelin antagonists with high ETA selectivity. Potency- and solubility-enhancing modifications. J Med Chem 35:2139–2142
84. Krystek SJ, Bassolino DA, Bruccoleri RE, Hunt JT, Porubcan MA, Wandler CF, Andersen NH (1992) Solution conformation of a cyclic pentapeptide endothelin antagonist. Comparison of structures obtained from constrained dynamics and conformational search. FEBS Lett 299: 255–261
85. Reily MD, Thanabal V, Omecinsky DO, Dunbar JJ, Doherty AM, DePue PL (1992) The solution structure of a cyclic endothelin antagonist, BQ-123, based on 1H-1H and 13C-1H three bond coupling constants. FEBS Lett 300:136–140

86. Satoh T, Barlow D (1992) Molecular modelling of the structures of endothelin antagonists. Identification of a possible structural determinant for ET-A receptor binding. FEBS Lett 310:83–87
87. Warner TD, Allcock GH, Corder R, Vane JR (1993) Use of the endothelin antagonists BQ-123 and PD 142893 to reveal three endothelin receptors mediating smooth muscle contraction and the release of EDRF. Br J Pharmacol 110:777–782
88. Sogabe K, Nirei H, Shoubo M, Nomoto A, Ao S, Notsu Y, Ono T (1993) Pharmacological profile of FR 139317, a novel, potent endothelin ETA receptor antagonist. J Pharmacol Exp Ther 264:1040–1046
89. Tschudi MR, Lüscher TF (1994) Characterization of contractile endothelin and angiotensin receptors in human resistance arteries: evidence for two endothelin and one angiotensin receptor. Biochem Biophys Res Commun 204:685–690
90. Haynes WG, Strachan EF, Webb DJ (1994) ETB-receptors mediate vasoconstriction in vivo in man (abstract). Hypertension 24:401
91. Haynes WG, Webb DJ (1994) Contribution of endogenous generation of endothelin-1 to basal vascular tone. Lancet 344:852–854
92. Nayler WG (1992) Amlodipine. Springer, Berlin Heidelberg New York
93. Clozel M, Breu V, Burri K, Cassal JM, Fischli W, Gray GA, Hirth G, Loffler BM, Muller M, Neidhart W et al (1993) Pathophysiological role of endothelin revealed by the first orally active endothelin receptor antagonist. Nature 365:759–761

Endothelial Dysfunction: Clinical Applicability of Calcium Antagonist Drugs

H. L. Elliott

Introduction

Hypertension and other cardiovascular risk factors (such as hypercholesterolaemia and diabetes mellitus) are known to be associated with endothelial dysfunction but, as yet, there is no specific treatment capable of restoring normal endothelial function. In the absence of specific therapies, however, it is possible to adopt the alternative strategy of administering established and effective antihypertensive agents which additionally may be capable of modifying or reducing the extent and consequences of underlying endothelial dysfunction. For example, the most obvious manifestation of endothelial dysfunction in hypertension is impairment of nitric oxide dependent vasodilator mechanisms and, as a consequence, a relative enhancement of the responses to vasoconstrictor substances such as noradrenaline, angiotensin II and endothelin 1. It is well recognised that calcium antagonist drugs are capable of attenuating the vasoconstrictor responses to all of these substances (Pasanisi et al. 1985; Kiowski et al. 1991; Ueda et al. 1993). An additional consideration is that calcium antagonist drugs have been shown to have anti-atherosclerotic properties which provide therapeutic potential for beneficially modifying a further manifestation of endothelial dysfunction. To date, however, there is no clear evidence in man that calcium antagonist drugs can control blood pressure, delay the development of atherosclerotic lesions and reduce cardiovascular morbidity and mortality. This current lack of evidence does not necessarily indicate that calcium antagonists are unable to provide such benefits in intact man but instead it may reflect the shortcomings and weaknesses of early treatment regimens with prototype calcium antagonist drugs.

Problems with Early Types of Dihydropyridine

Duration of Action

The prototype dihydropyridine, nifedipine, in its original capsule formulation is typically administered three or four times daily. While there is no doubt that nifedipine is an effective antihypertensive drug, the duration of action of this formulation is not only short but of widely variable extent. More importantly, such rapid changes and wide fluctuations in the vasodilator/antihypertensive activity of short – acting dihydropyridines lead to activation of the renin angiotensin and sympathetic nervous systems even during long-term treatment. For example, it has been shown with another first generation dihydropyridine, felodipine, that there are 50%–100% increases in circulating catecholamines and corresponding increases in heart rate even after one year of treatment when steady state conditions would have been expected (Leenen and Holliwell 1992). Thus, short-acting dihydropyridines may be very effective vasodilator antihypertensive drugs but it is highly probable that any potential, additional benefits of their calcium antagonist activity may be compromised, or totally lost, if neurohumoral mechanisms are repeatedly stimulated during long-term treatment. Since it has been shown that noradrenaline and angiotensin II potentiate the responses to endothelin 1, for example, these short-acting dihydropyridines are not well suited to significantly improving endothelial function as part of a long-term treatment strategy.

Studies of Atherosclerosis in Man

Endothelial dysfunction is an integral part of the atherosclerotic process which underlies the development of cardiovascular diseases. To date, there have been three major clinical trials investigating the effect of calcium antagonist treatment on the development and progression of atherosclerotic lesions. For example, the INTACT study using treatment with nifedipine (capsules) 20 mg q.i.d. for 3 years demonstrated significant reductions in the progression of atherosclerotic lesions (Lichtlen et al. 1990). These were small improvements of no immediate clinical relevance but of potential long-term importance in leading to a reduced rate of atherosclerotic vascular complications: similar, small improvements in respect of early lesions have also been reported in the other studies with nicardipine and with isradipine (Waters et al. 1990; MIDAS 1994). Thus, there is some evidence that the anti-atherosclerotic properties of calcium antagonist drugs are detectable in patients with vascular disease with beneficial effects particularly against the development and progression of early lesions (Table 1).

Unfortunately, these potentially beneficial effects on atherosclerotic lesions were totally overshadowed by the excess of morbid and mortal cardiovascular events in these three clinical trials (Table 1). However, all three of these drugs are of short dura-

Table 1. Calcium antagonists and atherosclerosis in humans

	Positive findings	Negative Results
Nicardipine (30 mg tid)	28% reduction in rate of development of new lesions; reduced rate of progression of minimal lesions	Acute myocardial infarction in 14 of 192 patients (compared to eight on placebo)
Nifedipine (capsule 20 mg qid)	Progression of lesions in 24% of patients (27% on placebo) and regression in 14% (10% on placebo)	Cardiac death in eight of 173 patients (compared to two on placebo)
Isradipine (2.5 mg bd)	Lesser increases in intima-medial thickness within first 6 months (preliminary analysis)	Increased angina in 11 patients (compared to three on hydrochlorthiazide)

tion of action and associated with reflex cardioacceleration and stimulation of the sympathetic nervous system. It is possible to conclude that the cardiovascular problems which occurred in the three clinical trials reflected the shortcomings of these particular drug and formulations rather than the intrinsic failure of the calcium antagonist effects of the drugs themselves.

The same rationale can be applied to the studies of the myocardial ischaemic syndromes in which the problems of reflex cardioacceleration and sympathetic stimulation are again the probable determinants of the deleterious effects obtained with short-acting dihydropyridines (Held et al. 1989).

The Profile of Future Dihydropyridine Drugs

To realise the potential benefits of calcium antagonism in relation to endothelial function (and in addition to blood pressure control) it is necessary to modify the drug delivery systems for short-acting dihydropyridines in order to provide fully and consistently maintained calcium antagonist activity and, equally importantly, to avoid provoking the activation of the renin angiotensin and sympathetic nervous systems.

Rate of Drug Delivery

The rate of drug delivery into the systemic circulation has a profound effect on the haemodynamic and neurohumoral responses to a dihydropyridine calcium antagonist drug. In an elegant study of intravenous nifedipine, comparing a rapid bolus dose and a slow infusion, it was clearly shown that the *rate* of drug delivery determined the pattern of response (Fig. 1a) (Kleinbloesem et al. 1987). Thus, the rapid attainment of an effective plasma concentration via a rapid intravenous bolus dose (and exponential infusion) caused no appreciable fall in blood pressure because there was an associated increase in adrenergic activity and an increase in heart rate (and cardiac output) (Fig. 1b). However, if an identical plasma nifedipine concentration was gradually attained over several hours by means of a progressive incremental intravenous infusion there was no adrenergic response and no increase in heart rate but there was a significant decrease in blood pressure (Fig. 1b).

Generally, and presumably as a result of "re-setting" of baroreflex responsiveness, there is less sympathetic activation during steady state treatment with a dihydropyridine calcium antagonist. However, if the drug or its formulation continue to cause wide fluctuations in drug concentrations and vasodilator effects during apparent steady state treatment then cardioacceleration will continually recur. Experimentally this was also clearly demonstrated in the intravenous infusion study when a further bolus dose of nifedipine was superimposed upon the steady state infusion (Fig. 1c). This further bolus led again to a clear increase in heart rate and attenuation of the hypotensive response (Fig. 2b). Several clinical studies have clearly demonstrated that sympathetic activation persists during long-term treatment when rapid-onset/short-acting dihydropyridines are administered (Ruzicka and Leenen 1995).

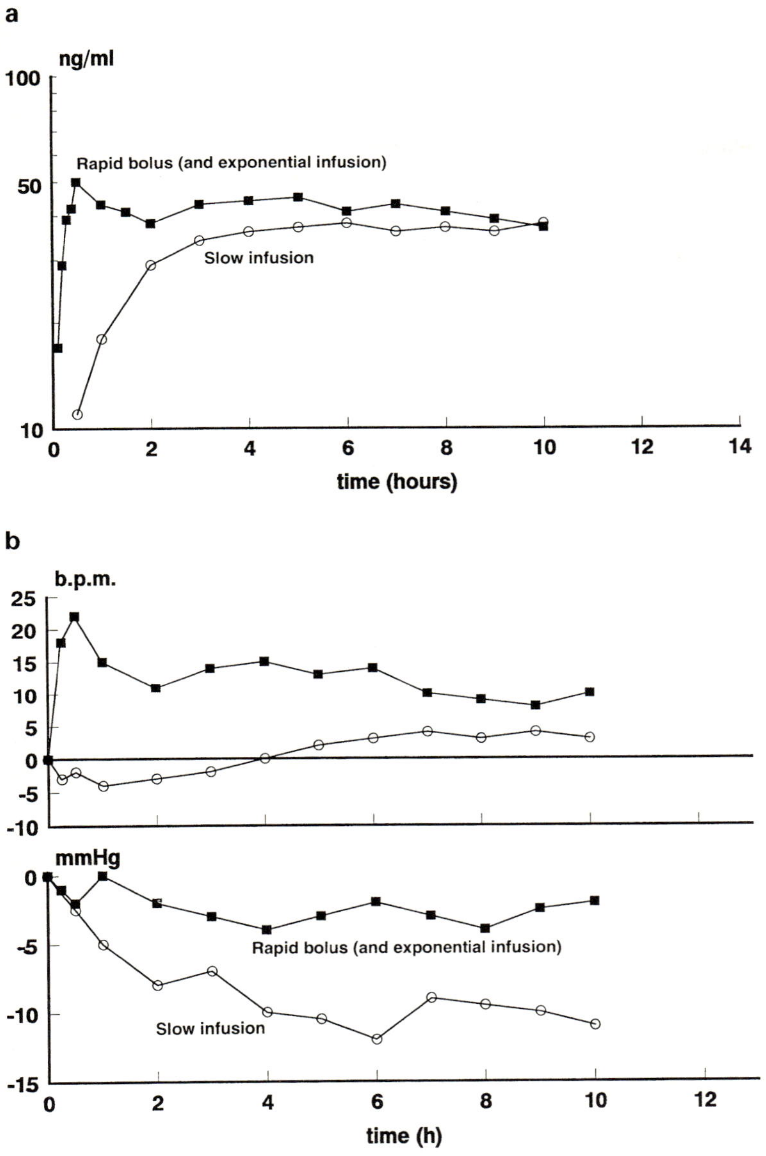

Fig. 1. a Comparison of the plasma nifedipine concentration–time profiles following rapid bolus injection and following slow infusion. **b** Comparison of the heart rate (*top*) and diastolic blood pressure responses (*bottom*) following bolus and infusion delivery of nifedipine

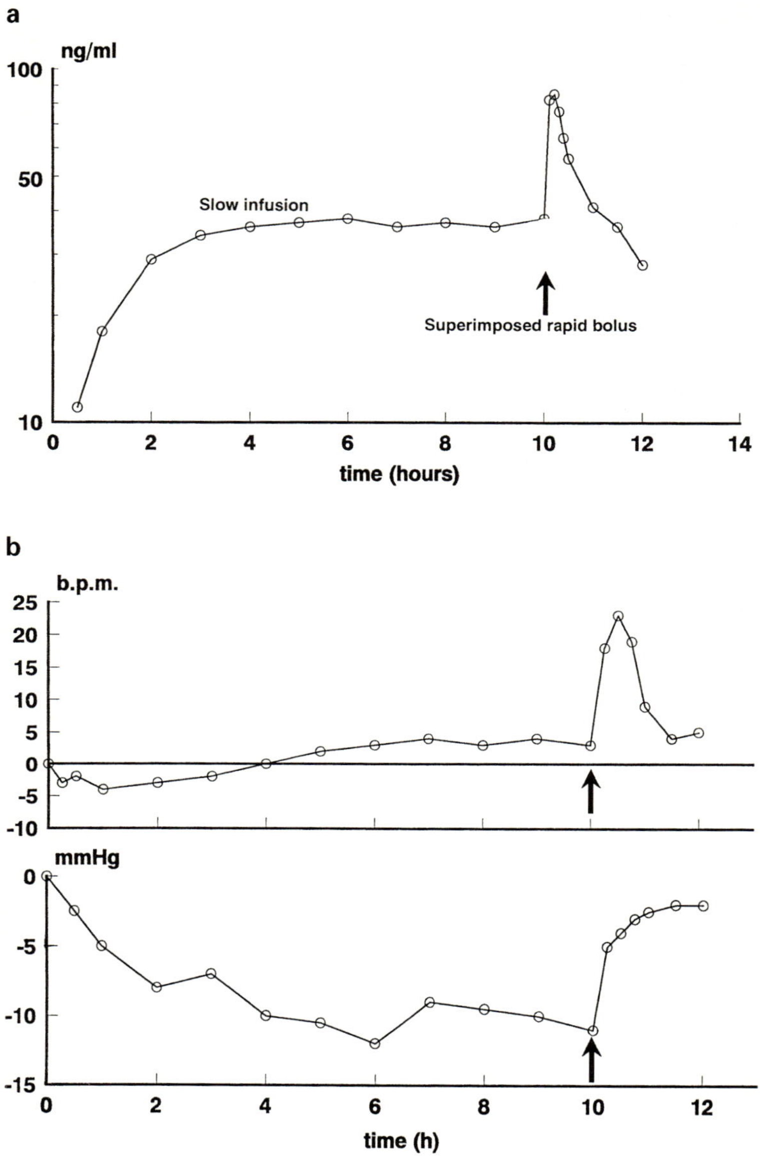

Fig. 2. **a** Plasma concentration–time profile following a rapid bolus dose of nifedipine superimposed upon the slow infusion. **b** Heart rate (*top*) and diastolic blood pressure responses (*bottom*) to the superimposed bolus dose of nifedipine

Long-Acting Dihyropyridines

The rate of delivery of a dihydropyridine calcium antagonist drug and the avoidance of wide fluctuations in vasodilator activity is the crucial first step for the avoidance of sympathetic activation and for the possible attainment of the additional therapeutic benefits of calcium antagonism. However, it is also necessary that the pharmacological activity is fully and consistently maintained during long-term treatment. It has been clearly established that there is a direct relationship between the plasma concentration time profiles for calcium antagonist drugs (e.g. nifedipine, felodipine and amlodipine) and the time profiles for the reduction of blood pressure (Donnelly et al. 1988, 1993; Blychert et al. 1992) and for the attenuation of vascular reactivity (Ueda et al. 1993; Bainbridge et al. 1993). There is an obvious requirement therefore for a drug concentration profile which is consistently maintained throughout 24 h because this will lead to consistent pharmacological activity throughout 24 h. In this respect there is a remarkable and coincidental similarity between the steady state plasma level of about 40 nmol/l which was achieved with the incremental intravenous infusion of nifedipine (as described by Kleinbloesem) and the corresponding plasma concentration time profile for the delivery of nifedipine from the gastrointestinal therapeutic system (GITS) formulation (Chung et al. 1987) (Fig. 3). Thus, over a period of approximately 6 h after the first dose of nifedipine GITS, there is a gradual attainment of the effective therapeutic concentration and this concentration is then maintained consistently throughout the full 24 h. This GITS delivery system is able to reproduce a desirable, gradual onset profile and sustained drug levels throughout 24 h in contrast to the more rapid onset profiles and short durations of the capsule (and tablet) formulations of nifedipine.

There are not yet long-term studies of nifedipine GITS directly investigating its effect on endothelial function but there is interesting preliminary evidence from stu-

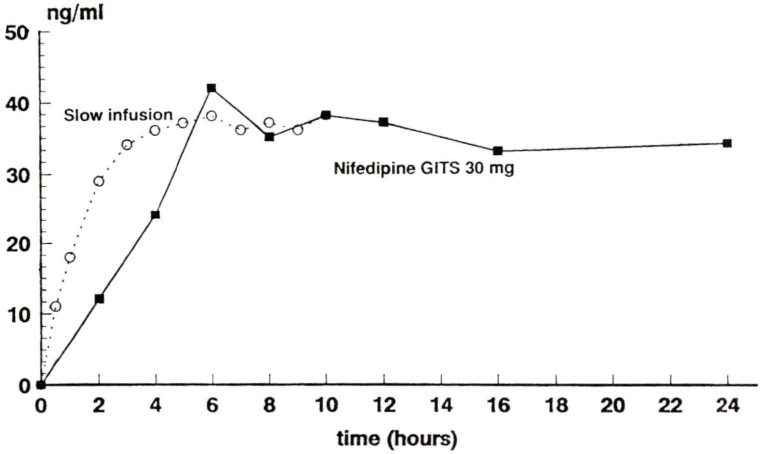

Fig. 3. Comparison of the plasma concentration–time profiles for nifedipine delivered by slow intravenous infusion and orally via the GITS formulation

dies of microalbuminuria which might be regarded as a surrogate index of endothelial dysfunction. In clinical trials comparing enalapril either with nifedipine GITS or with another first generation, short-acting dihydropyridine, nitrendipine, there were contrasting results. With nitrendipine the reduction in urinary albumin excretion was trivial in comparison to the effect of enalapril (Bianchi et al. 1991). At first sight, this might be interpreted as evidence that enalapril is a more effective therapy for reducing microalbuminuria (and, by implication, for reversing this particular manifestation of endothelial dysfunction). However, the concentration–effect profile for nitrendipine is very variable with an obvious peak plasma level at about 2–4 h post-dose and minimal drug levels at the end of the dosage interval: this obviously means that nitrendipine is producing minimal pharmacological activity at the end of its dosage interval after having produced a pronounced early peak response at about 2–4 h post-dose. This failure to produce a consistent effect throughout the dosage interval is again indicative of the potential problems of agents which have sub-optimal pharmacokinetic and haemodynamic profiles. In contrast, in a study comparing nifedipine GITS and enalapril there was an exactly comparable 37% reduction in microalbuminuria (De Quattro et al. 1995). Whilst these results require confirmation through further clinical trials, the findings are entirely consistent with the concept that there are benefits to be obtained with long-acting drugs which produce consistent effects that are not compromised by sub-optimal haemodynamic profiles and by neurohumoral activation.

A further indication that fully maintained haemodynamic and antihypertensive effects may be relevant in microalbuminuria is found in a study of "dipper" and "non-dipper" hypertensive patients (Bianchi et al. 1994). The level of overnight blood pressure and the absence of an overnight "dip" was shown to correlate with the degree of microalbuminuria such that "non-dippers" showed higher rates of microalbuminuria. The therapeutic requirement for reducing overnight blood pressure and, by implication, reducing microoalbuminuria therefore demands that the antihypertensive treatment should maintain its efficacy throughout the night and this again is an argument in favour of a dihyropyridine calcium antagonist with a protracted duration of action.

Conclusions

With first generation dihydropyridine drugs and formulations, the fluctuations in plasma drug concentrations and the corresponding fluctuations in calcium antagonist activity are not capable of providing a consistent and sustained therapeutic effect during long-term treatment. In addition, continuing or recurring neurohumoral stimulation is likely to substantially diminish the potential benefits of treatment. To provide optimal blood pressure control and to realise the potential benefits in respect of endothelial function it seems logical that calcium antagonist activity should be consistently present throughout the dosage interval, throughout the whole 24-h period, and throughout the period of long-term treatment. Whilst this may be possible with an intrinsically long-acting dihydropyridine derivative, such as amlodipine, it is necessary with short-acting dihydropyridines to utilise a sophisticated drug delivery system, such as the GITS formulation of nifedipine, which provide fully sustained profiles of calcium antagonist activity without the deleterious effects of neurohumoral activation.

Abstract

Endothelial dysfunction is now known to be an early feature of atherosclerotic cardiovascular diseae and, with antihypertensive treatment, it may not be possible to obtain maximum long-term prevention unless the treatment has a beneficial effect on endothelial function in addition to blood pressure control.

The anti-atherosclerotic potential of calcium antagonist drugs is well established but, in humans, the potentially beneficial long-term anti-atherosclerotic effects have been completely negated by the excess of cardiac events. However, the drugs used in these clinical trials are known to be of rapid onset and short duration of action, with a recurring ability to provoke activation of the sympathetic and renin angiotensin systems. These undesirable neurohumoral responses can be effectively abolished by means of a drug delivery system which provides consistent pharmacological activity throughout 24 h. Thus, the long-acting dihydropyridines are better suited to providing the optimal therapeutic approach with fully sustained 24-h blood pressure control without reflex sympathetic activation.

References

Bainbridge AD, Herlihy O, Meredith PA, Elliott HL (1993) A comparative assessment of amlodipine and felodipine ER: pharmacokinetic and pharmacodynamic indices. Eur J Clin Pharmacol 45:425–430

Bianchi S, Bigazzi R, Galdari G, Campese VM (1991) Microalbuminuria in patients with essential hypertension. Am J Hypertens 4:291–296

Bianchi S, Bigazzi R, Baldari G, Gianpaolo Sgherri, Campese VM (1994) Diurnal variations of blood pressure and microalbuminuria in essential hypertension. Am J Hypertens 7:23–29

Blychert E, Edgar B, Elmfeldt D, Hedner T (1992) Plasma concentration-effect relationships for felodipine: a meta analysis. Clin Pharmacol Ther 52:80–89

Chung M, Reitberg DP, Gaffney M, Singleton W (1987) Clinical pharmacokinetics of nifedipine gastrointestinal therapeutic system. Am J Med 83 [Suppl 6B]:10–14

Donnelly R, Elliott HL, Meredith PA, Reid JL (1988) Nifedipine: individual responses and concentration-effect relationships. Hypertension 12:443–449

Donnelly R, Meredith PA, Howie CA, Elliott HL (1993) Pharmacodynamic modelling of the antihypertensive response to amlodipine. Clin Pharmacol Ther 54:303–310

Held PH, Yusuf S, Furberg CD (1989) Calcium channel blockers in acute myocardial infarction and unstable angina: an overview. Br Med J 299:1187–1192

Kiowski W, Lüscher TF, Linder L, Buhler FR (1991) Endothelin 1-induced vasoconstriction in humans: reversal by calcium channel blockade but not by nitrovasodilators or endothelium-derived relaxing factor. Circulation 83:469–475

Kleinbloesem CH, van Brummelen P, Danhof M, Faber H, Urquhart J, Breimer DD (1987) Rate of increase in the plasma concentration of nifedipine as a major determinant of its haemodynamic effects in humans. Clin Pharmacol Ther 41:26–30

Leenen FHH, Holliwell DL (1992) Antihypertensive effect of felodipine associated with persistent sympathetic activation and minimal regression of left ventricular hypertrophy. Am J Cardiol 69: 639–645

Lichtlen PR, Hugenholtz P, Rafflenbeul W, INTACT group (1990) Retardation of the angiographic progression of coronary artery disease in man by the calcium channel blocker nifedipine. Results of the international nifedipine trial on antiatherosclerotic therapy (INTACT). Lancet 335 (8698):1109–1113

MIDAS (1994) Multicentre Isradipine Diuretic Atherosclerosis Study: 15th Scientific Meeting of the International Society of Hypertension

Pasanisi F, Elliott HL, Meredith PA, Sumner DJ, Reid JL (1985) Effect of calcium channel blockers on adrenergic and non-adrenergic vascular response in man. J Cardiovasc Pharmacol 7:1166–1170

Quattro V de, Lee D-P, Lu ZW, Wang LD (1995) Improvement in early morning blood pressure control and through: peak ratio with equivalent reduction of proteinuria after nifedipine GITS vs enalapril for hypertension. (in press)

Ruzicka M, Leenen FHH (1995) Relevance of intermittent increases in sympathetic activity for adverse outcome on short-acting calcium antagonists. In: Laragh JH, Brenner BM (eds) Hypertension: pathophysiology, diagnosis and management, 2nd edn. Raven, New York, pp 2815–2825

Ueda S, Meredith PA, Howie CA, Elliott HL (1993) A comparative assessment of the duration of action of amlodipine and nifedipine GITS in normotensive subjects. Br J Clin Pharmacol 36:561–566

Waters D, Lesperance J, Francetich M et al (1990) A controlled clinical trial to assess the effect of a calcium channel blocker on the progression of coronary atherosclerosis. Circulation 82:1940–1953

Endothelin in Hypertension and Diabetes Mellitus

A. Fernandez-Cruz, R. Patiño, J. Ibarra, L. Fernandez, M. Roldan, J. Moya, and R. Fernandez-Durango

Diabetes and hypertension are two of the most prevalent chronic diseases afflicting mankind. Individuals with one of these maladies often suffer from the other and it is now increasingly recognized that such concurrence is greater than what would be predicted by chance alone.

Diabetes mellitus is associated with high cardiovascular morbidity and mortality. As is usual with this disease, in which the complexities abound and the categorization is crucial, we easily rush into generalities rather than consider subdivisions, as for instance between type I and type II diabetes.

Patients with type I (insulin-dependent) diabetes (IDD), who are prone to diabetic nephropathy, have a familial predisposition to develop essential hypertension [1].

Furthermore, the normotensive offspring of non-diabetic essential hypertensives tend to have high plasma insulin levels and reduced insulin sensitivity; these individuals are therefore prone to develop type II diabetes. Indeed, essential hypertension has been called an "insulin-resistant state" [2–4]. Nevertheless, not all patients with diabetes also develop essential hypertension, and, conversely, not all essential hypertensives develop diabetes; thus, it is as important to establish the basis for these differences in expression of the two diseases as it is to determine the underlying causes of their overlapping features.

Blood Pressure in Insulin-Dependent Diabetes

Recent studies suggest that there is no marked difference in blood pressure among young uncomplicated insulin-dependent diabetics when compared with their siblings [5], the only difference being a 2- to 3-mm increase in diastolic pressure, which is of theoretical rather than practical importance. Tarn et al. [5] have shown similar inheritance of blood pressure form parents to diabetic and non-diabetic children, and with the previously described correlates, this suggests that the uncomplicated hyperglycemic state does modify neither genetic, nor environmental influences on blood pressure. lt has become clear that much of the hypertension in IDD patients is related to the presence of diabetic renal disease, even at an early stage. Most studies, although retrospective, have shown no difference in blood pressure between subjects who do, and do not, develop nephropathy prior to its development. A detailed prospective study by Mathielson et al. clearly shows microalbuminuria to occur before blood pressure rises [6]. Furthermore, Norgaard et al. indicated in their survey that the prevalence of hypertension among the IDD and non-diabetic Danish population is virtually identical, once all IDD patients with albumin excretion rates over 30 mg/day are excluded [7].

The precise pathogenesis of human diabetic kidney disease and the factors responsible for the susceptibility to it remain to be established. However, there is now evidence that renal disease clusters in families and that genetic factors are of central importance in determining liability.

A predisposition to arterial hypertension has been suggested as playing a contributory role in the development of kidney disease. A familial transmission of increased blood pressure through increased lithium/sodium countertransport-mediating arterial wall damage and glomerular injury is still a matter of debate [8].

Based on the observations available one can hypothesize that genetically controlled hypertrophic processes may be implicated in the susceptibility to arterial wall damage and glomerular injury in type I diabetes. Increased fibroblast Na^+/H^+ exchange has been reported in patients with diabetic nephropathy [8]. Further, young African-Americans with borderline hypertension and insulin resistance have been shown to have elevated erythrocyte membrane N^+/H^+ exchange activity in comparison to matched normotensive and hypertensive subjects without insulin resistance. The first signs of renal damage are indeed the appearance of microalbuminuria and of a small elevation in arterial pressure, changes associated with significant mesangial expansion. Microalbuminuria is associated with abnormalities of lipoprotein profiles, possibly as a consequence of insulin resistance-induced hyperinsulinemia.

Nearly all studies confirm that, in IDD, hypertension is usually only a symptom of underlying diabetic renal disease. In other words, a disruption in the circadian blood pressure pattern may help to identify those patients with type I diabetes who will eventually develop renal disease. Essential hypertension is probably as common among IDD patients as in the general population; however, in response to the genetically controlled injury of the arterial wall and glomerular initiation and mediation of subintimal changes it can lead to the development on endothelial dysfunction. The endothelial dysfunction will then express itself as an impaired release of nitric oxide that, in the presence of tromboxanes angiotensin II and endothelin, will increase peripheral vascular resistance. An additional factor that may play a role in the maintenance of the increased peripheral vascular resistance observed in the hypertensive IDD patients is the development of an insulin-resistant state.

On the other hand, evidence that salt retention and transient volume expansion may play a key role in the pathogenesis of at least some essential hypertensives has been intensively investigated. Numerous investigators have also obtained evidence for possible abnormalities of salt and water metabolism in both type I and type II diabetes, as well as in experimental diabetes. One of the characteristic findings in both type I and type II diabetics without nephropathy is a significant increase in exchangeable body Na^+ [9]. This Na^+ retention is associated with an expansion of extracellular fluid volume [10]. Furthermore, the key role of the kidneys in hypertension is also widely recognized. Renal parenchymal dysfunction with impaired sodium excretion invariably leads to hypertension. These findings are consistent with the view that the primary problem in insulin-dependent hypertensive diabetics is an impaired renal function which, because of genetic or acquired defects, is unable to handle the salt load with which the kidneys are presented. This results in retention of Na^+, Cl^-, and water, and expansion of plasma volume. Guyton [11] has emphasized the fact that the blood pressure–natriuresis relationship indicates that the kidneys may have a very large capacity to compensate for plasma volume expansion by mediating a pressure induced natriuresis. Atrial natriuretic peptide (ANP) and the renin–angiotensin system are reciprocally related to the control of extracellular volume and sodium excretion in man. A significant diversity of opinion exists in the literature about the status of the renin–angiotensin system and the plasma levels of ANP in diabetics. In view of the complexities of volume control in diabetes differences in renin and ANP are not altogether surprising.

Our results [12], along with others from the literature, do not support the hypothesis that the ANP is a main determinant of the increased exchangeable body sodium in type I patients with incipient nephropathy.

Alternatively, the idea that the salt and water retention may be the consequence of elevated plasma ouabain levels in diabetes has also been considered [12]. Recent studies provide direct evidence that ouabain can cause hypertension. Parenteral administration of ouabain to normal rats for a month or more causes the blood pressure to rise progressively [13]; the blood pressure can then be lowered by stopping the ouabain treatment or by injecting anti-ouabain antibodies [14]. In humans a new clinical syndrome has been identified, a "primary ouabainoma" [15]; in this case, the patient's plasma ouabain levels and blood pressure both declined after surgical removal of the tumorous adrenal gland.

In summary, there is now strong evidence that ouabain is a hypertensiogenic hormone. It therefore seems appropriate to determine precisely how ouabain might be involved in the pathogenesis of type I hypertensive diabetics.

Blood Pressure in Non-Insulin-Dependent Diabetes

Controversy has raged for over 50 years as to whether hypertension is associated with non-insulin-dependent diabetes (NIDD). Type II diabetes mellitus is however associated with peripheral insulin resistance and, in the great majority of cases, with obesity. The association of hypertension and obesity has been recognized for decades. Building on observations made by the French clinician Jean Vague decades ago [16], recent large scale epidemiologic studies have clearly demonstrated that the pattern of body fat distribution is related to the cardiovascular and metabolic complications of obesity. Upper body segment obesity, particularly the accumulation of intraabdominal fat, is associated with hypertension, ischemic heart disease, and type II diabetes mellitus.

The concept of the insulin resistance syndrome has been extensively discussed elsewhere and will not be repeated here. The current popular view is that hyperinsulinemia is a consequence of insulin resistance insofar as increased insulin secretion is required to maintain normal or near-normal glucose uptake in skeletal muscle. Elevated insulin levels and hypertension, therefore, are both associated with abdominal obesity. Recent studies, moreover, indicate that insulin levels are directly related to blood pressure among both the obese and, interestingly, the non-obese as well [17, 18]. There are many effects of insulin which potentially could induce hypertension:

1. Hyperinsulinemia seen in obese subjects is responsible for the increased activity of the sympathetic nervous system.
2. Insulin stimulates sodium potassium ATPase activity directly resulting in an increase in distal tubular sodium reabsorption.
3. The effect of insulin on erythrocyte sodium potassium ATPase in hypertension is impaired.
4. Hyperinsulinemia could act as a potent stimulus to the growth of vascular smooth muscle cells, thus leading to structural changes in the vascular bed.

One of the interesting and unanticipated observations to emerge in this area is that insulin resistance is frequently present in non-obese hypertensives. How does the hypothesis outlined in previous sentences accommodate with this observations? Consider a young individual with well-developed thermogenic mechanisms. Such an individual, by overeating or consuming a high fat diet, might develop insulin resistance. Insulin resistance, parenthetically, occurs rapidly in animals placed on a high-fat diet [19]. The consequent hyperinsulinemia and sympathetic stimulation would, by driving efficient thermogenic mechanisms, compensate for the increased caloric intake and maintain energy balance by increasing metabolic rate. The price paid for resisting obesity is insulin resistance, hyperinsulinemia and hypertension. It has been suggested then that either excess insulin action and/or peripheral insulin resistance may, separately or together, be important factors explaining both the increased vascular resistance and the expanded extracellular volume (increased exchangeable sodium) seen in diabetic hypertension. Accordingly, the hyperinsulinemia associated with type II diabetes and overtreated type I diabetes could cause high blood pressure by enhancing the sympathetic nervous system (SNS) activity. However, the role of insulin in the hypertensive process has been questioned. Dr. Hall and his colleagues from the University of Mississippi tested the hypothesis in dog and rat models [20] of hypertension. The studies in dogs failed to support the idea that hyperinsulinemia causes hypertension or potentiates the hypertensive action of these pressor systems. In three rat models they found that hyperinsulinemia elicited a chronic increase in blood pressure, a mild hypertension. The logical question is whether the dog or the rat model mimics the response in humans more closely. Further acute insulin infusion in normal human volunteers that produced physiologic hyperinsulinemia decreased peripheral vascular resistance despite increasing sympathetic nerve activity [21]. This finding led us to conclude that a critical assumption of the insulin hypothesis, i.e., that insulin increases vascular resistance (and arterial pressure) since it increases SNS activity, is not valid.

In lean, normotensive individuals insulin-induced vasodilation (a depressor action, mediated by nitric oxide) offsets the four pressor effects of insulin. But what effect would insulin infusion have on obese insulin-resistant, hypertensive patients? At the University of lowa Dr. Mark showed, to his surprise, that "substantial increases" in plasma insulin did not affect arterial blood pressure "in either lean normotensives or the obese, insulin-resistant, mildly hypertensive subjects." Thus, hyperinsulinemia per se does not acutely cause hypertension in spite of its short term effects on SNS activity and renal sodium reabsorption.

However, other evidence suggests that long-term sustained hyperinsulinemia may indeed play an etiological role in the development of hypertension in only certain subgroups of type II diabetes [22].

A second alternative is to consider that insulin may promote hypertension via its effects on vascular remodeling and the atherosclerotic processes. Furthermore recent studies nave shown that the vasodilator effects of insulin may not exist in individuals with a insulin resistance [23, 24], and that insulin resistance may even be associated with paradoxical vasoconstrictive response to insulin. Lastly, there is accumulating evidence that either too little insulin or the decreased effective insulin action associated with insulin resistance may contribute to enhanced vascular tone. Thus in insulin resistant states such as type II diabetes obesity and essential hypertension, insulin may exert both acute and chronic hypertension promoting effects.

The third possibility explaining the association of insulin resistance and hypertension may have a genetic basis, namely that these two develop in parallel and independently from each other, shared or controlled by two separate gene loci. In any case, a genetic background to NIDD behaves like a risk factor, not only for diabetes, but also for hypertension.

In both normotensive and hypertensive metabolically stable type II NIDD patients, an increased exchangeable body sodium has been reported. Our own data [12] and others in the literature show an increase in plasma ANP concentrations in type II diabetics, associated with a blunted increase following volume expansion.

There is accumulating evidence for multiple interactions between ouabain-like factors (OLF) and ANP systems [25].

The potential role of endogenous natriuretic factors and ANP in type II hypertensive diabetics is a matter of speculation and needs further investigation. However, as we mentioned above, exchangeable body sodium is increased in diabetics an even more in patients with established nephropathy. Thus the possible volume-induced elevation of endogenous OLF make them excellent candidates for mediating cellular ion handling that could result in increased vascular smooth muscle contraction observed in diabetic individuals with hypertension [25]. Moreover, the blunted ANP response and the suggested renal resistance to the action of the peptide may reconcile previous apparently divergent observations by various groups of investigators.

Lastly, diabetic individuals have been reported to frequently have extracellular and intracellular magnesium deficiency. This could result in increased intracellular Ca^{+} levels, which, in turn, would enhance vascular contractility and impair cellular insulin action, all of which are characteristics of type II diabetes associated with hypertension. This concept needs additional testing.

Since the discovery of endothelium-dependent vasodilation by Furchgott and colleagues in 1980 it has become evident that endothelial cells play a key role in the motor effects of certain vasoactive substances [26]. One of the characteristics of diseased vascular wall is the impairment of endothelium-dependent relaxation. This has been demonstrated repeatedly by several groups in animals with hypertension, in animals with atherosclerosis and in cerebral vasospasm. In contrast to endothelium-dependent relaxation, endothelium-depended contraction becomes prominent under acute pathologic conditions (e.g., hypoxia or reperfusion injury) or chronic conditions (e.g., hypertension, cerebral vasospasm or atherosclerosis).

Some studies show that the release of nitric oxide is impaired in experimentally induced diabetes mellitus [27, 28]. On the other hand, endothelin may have a role in hypertensive patients with type II diabetes, since it is a potent vasoconstrictor and because of its ability to stimulate the release of aldosterone [29], catecholamines [30], and renin.

Taken together, it can be speculated that nitric oxide and endothelin 1 may be involved in the pathogenesis of some forms of hypertension in type II diabetes mellitus, by a mechanism ultimately involving activation of voltage-dependent Ca^{2+} channels in human vascular smooth muscle.

Endothelin in Hypertension and Diabetes

Endothelin (ET) is synthesized, among others, by endothelial cells, macrophages and vascular smooth muscle cells, and constitutes the most potent endogenous vasoconstrictor known nowadays [29]. Accumulating evidence indicates that ET has a potent vasocontricting action in vitro and a pressor action in vivo [30]. It may therefore play a part in the control of blood pressure and the pathophysiology of hypertension. Recent studies using endothelin-specific antibodies suggest that ET plays an important role in the modulation of systemic blood pressure and renal function in spontaneously hypertensive rats [32–34].

Endothelin is present in plasma, perhaps as result of spillover from the interface of endothelial cell injury, as well as in essential hypertension [35]. Indeed the ET levels in essential hypertension is controversial, since both increased and also (mainly) normal circulating plasma levels have been reported [36–42]. In patients with essential hypertension, age-related changes in the renin–angiotensin–aldosterone system are well documented. Our findings convincingly indicate an important effect of age on the renin–angiotensin system as well as the ET levels, particularly alter age 50 [43]. Furthermore, our data agree with increased plasma levels in young essential hypertensives, but not in the elderly [44].

On the other hand, elevated plasma ET levels have been demonstrated in diabetes [45]. Kawamura et al. recently reported elevated plasma ET levels in type II patients with retinopathy [46]. In addition, increased plasma and urinary ET excretion in type II patients, with microalbuminuria and macroalbuminuria, was also found [47, 48]. Indeed, atheroscierosis is associated with increased plasma ET levels [49]. We obtained similar findings in the type II diabetics with macrovascular disease [50]. In diabetes mellitus elevated endothelin levels have been reported to act as a marker of hypertension [51].

Because of the current knowledge on endothelial dysfunction in hypertension and diabetes, which could involve alterations in the production and release of endothelium-derived relaxing as well as contracting factors we decided, in order to gain insight into the potential role of endothelin, to evaluate the plasma level of ET-1 in hypertensives diabetics.

Material and Methods

We have measured plasma ET in 191 diabetic patients: 50 type I (age, 33.6 ± 11 years; range, 17–65 years; mean duration, 15.6 years; range, 4–40 years), and in 141 type II (age, 62 ± 8 years; range, 36–81 years; mean duration, 14 ± 8 years; range, 2–35 years), and in 100 young (20–50 years) and 36 older (55–78 years) healthy control subjects. The identified diabetic complications were, for type I: hypertension, six patients; isolated retinopathy, 11 patients; retinopathy with associated microalbuminuria with creatinine < 1.5 mg, 16 patients; macrovascular disease, two patients; uncomplicated, 23 patients. Type II: hypertension, 71 patients; retinopathy, 52 patients; microalbuminuria with creatinine < 1.5 mg, 30 patients; macroalbuminuria with creatinine < 1.5 mg, 16 patients, macrovascular disease, 36 patients; uncomplicated, 37 patients.

Blood samples for ET-1 measurements were collected after 30 min seated rest, preserved in ice-cold ethylenediaminetetraacetic acid (EDTA)-coated polystyrene tubes and stored at −70°C for up to 2 months. ET levels were measured by a previously described radioimmunoassay [52]. The data are expressed as means ± SD.

Results

Our results show that endothelin levels increase with age in normal subjects (1.92 ± 0.53 pg/ml in the young versus 4.38 ± 1.8 pg/ml in the elderly, $p < 0.01$). Plasma ET-1 was significantly increased in uncomplicated type II diabetics, 5.9 ± 1.6 versus 4.38 ± 1.8 pg/ml in age-matched controls, $p < 0.005$.

However, the type I normotensive normoalbuminuric patients without macrovascular disease show endothelin levels which are considered as within the normal ranges, 1.9 ± 0.4 pg/ml. Furthermore, the results of plasma ET-1 in type I diabetics with isolated retinopathy (2.06 ± 0.36 pg/ml) were not significantly elevated versus control subjects (1.92 ± 0.5 pg/ml). On the other hand, type I albuminuric patients had a higher and significant elevation of plasma endothelin 1 than that of both normal controls and normotensive normoalbuminuric patients, 3.47 ± 0.75, $p < 0.001$ (Fig. 1).

No significant difference was noted either between type II diabetics with retinopathy without microalbuminuria hypertension or macrovascular disease and uncomplicated type II diabetics (5.4 ± 1.7 versus 5.9 ± 1.6 pg/ml). However ET-1 in type II diabetics with retinopathy and associated macrovascular disease were found to be significantly elevated versus uncomplicated patients (7.4 ± 2 versus 5.9 ± 1.6 pg/ml, $p < 0.05$). Once again, when those results are compared with the normal controls, the ET-1 plasma values were both found to be significant, $p < 0.05$ (Fig. 2). Because of the duration of the disease in our population among type II diabetics with retinopathy and the sample size of our investigation we can not support that plasma endothelin will play a major role in the development of diabetic retinopathy.

No significant difference was noted in the plasma levels of ET-1 in type II diabetics with abnormal albumin excretion, without macrovascular disease, when compared with uncomplicated patients (5.9 ± 1.7 versus 5.9 ± 1.6 pg/ml). However, when those

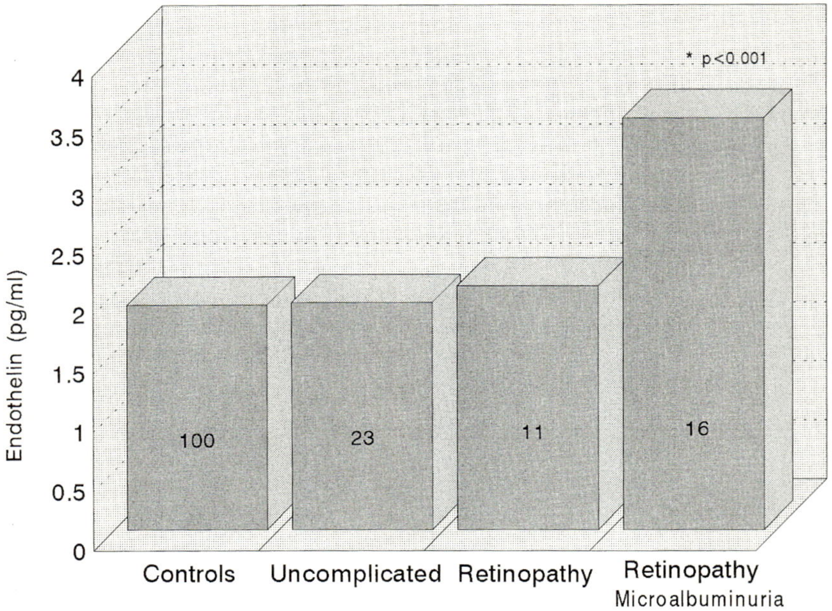

Fig. 1. Plasma endothelin-1 in type I diabetics, both normotensives with simple and proliferative retinopathy and microalbuminuric patients (>30 μg/min; <2 mg creatinine/dl). Numbers on the bars indicate patient populations

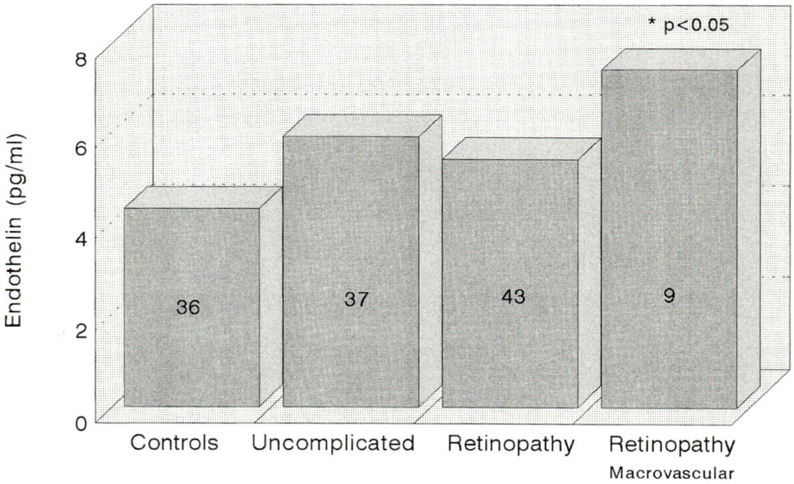

Fig. 2. Plasma endothelin-1 in type II uncomplicated diabetics and in normotensive normoalbuminurics with retinopathy with and without macrovascular disease. Numbers on the bars illustrate the patient populations

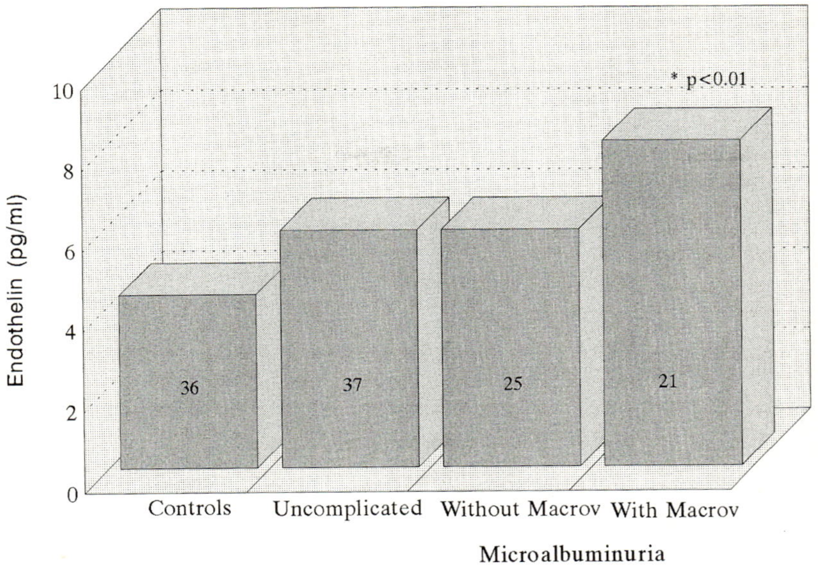

Fig. 3. Plasma endothelin-1 in type II uncomplicated and microalbuminuric patients (range, 20–200 μg/min; <2 mg creatinine/dl) with and without macrovascular disease. Numbers on the bars illustrate the patient populations

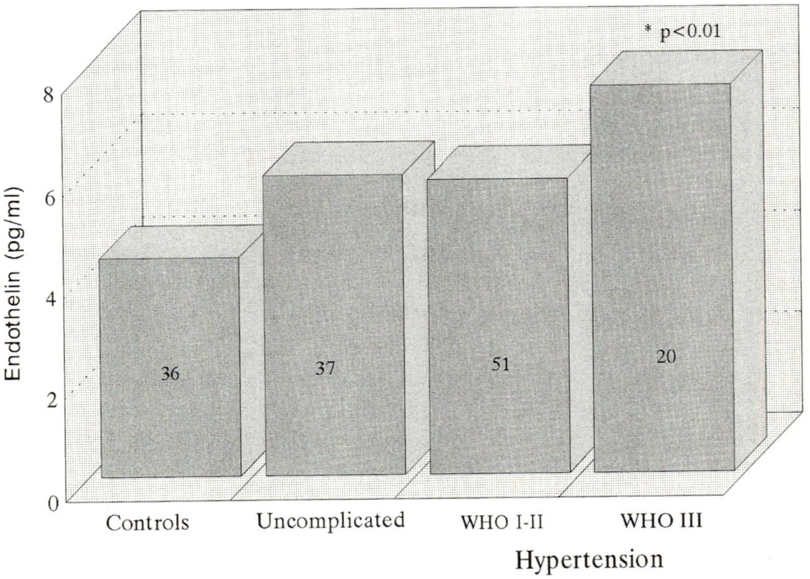

Fig. 4. Plasma endothelin-1 in type II uncomplicated normotensive and hypertensive diabetics in WHO stages I, II, and III. Numbers on the bars illustrate patient populations

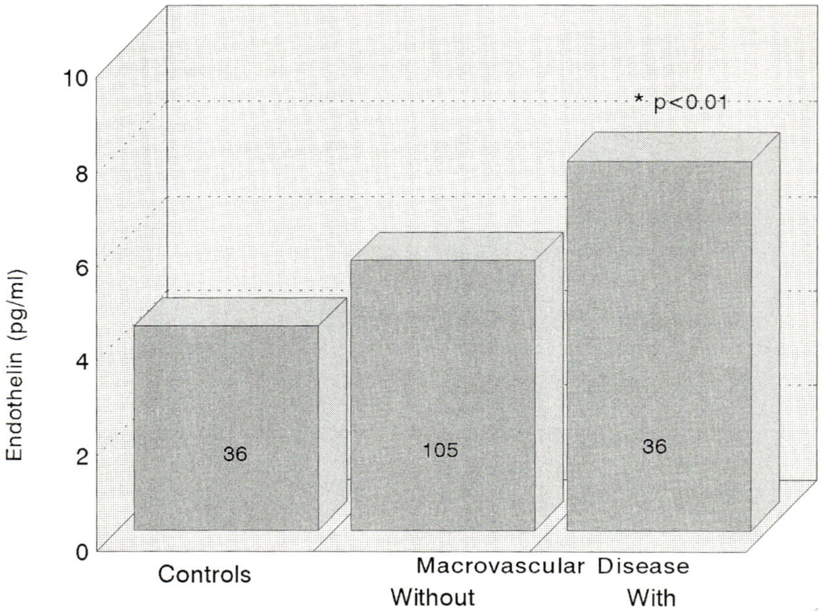

Fig. 5. Plasma endothelin-1 in type II diabetics with and without macrovascular disease (peripheral and coronary artery disease). Numbers on the bars illustrate patient populations

patients where considered in association with macrovascular disease plasma ET-1 values were found to be significantly elevated (8.1 ± 1.2 versus 5.9 ± 1.7 pg/ml, $p < 0.001$; Fig. 3).

When hypertension was investigated, the results continued to indicate that it is macrovascular disease which is inducing endothelin. Type II diabetic patients with hypertension in WHO stage I and II without macrovascular disease had endothelin levels which were not significantly different from those of the control group (5.8 ± 1.7 versus 5.9 ± 1.6 pg/ml). On the other hand the group of type II hypertensive diabetics in WHO stage III with macrovascular disease, showed significantly elevated ET levels (7.6 ± 1.2 pg/ml) when compared with either uncomplicated or hypertensives in WHO stages I and II ($p < 0.01$) (Fig. 4).

When macrovascular disease was considered (peripheral and coronary artery disease) in type II diabetic patients, the plasma ET levels were found to be significantly elevated as compared with those of type II diabetic patients without vascular complications (7.8 ± 2.4 versus 5.9 ± 1.7 pg/ml, $p < 0.01$) (Fig. 5).

Conclusion

ET levels were not increased in uncomplicated type I diabetes mellitus. However, they were increased in type II uncomplicated diabetes mellitus with an average of 15 years duration of the disease. We also found that the plasma levels of ET in simple and proliferative retinopathy in type I and type II diabetics, without macrovascular disease, were not significantly higher than those in normotensive uncomplicated diabetics. In addition when either microalbuminuria in type I or macrovascular disease in type II was considered our results clearly indicated a significant elevation of ET-1 plasma levels. Furthermore, ET-1 plasma levels in type II diabetics with abnormal albumin excretion and without macrovascular disease were found to be not significantly increased compared to uncomplicated diabetics, but they were significantly elevated versus age-matched control subjects.

Plasma ET levels in hypertensive type II diabetics in stages I and II of the WHO classification were found to be significantly elevated versus age-matched controls but normal in comparison to uncomplicated patients. On the other hand, when macrovascular disease is considered in WHO stage III hypertensives the plasma endothelin levels were found to be significantly elevated as compared with both normotensives and hypertensive diabetic patients without vascular complications. These observations suggest that the elevated plasma level of ET-1 immunoreactivity in patients with hypertension and type II diabetes is related to vascular complications, including atherosclerosis. There is now evidence that atherosclerosis is characterized by perturbations in endothelial vasomotor function. Moreover, emerging evidence suggests that increased ET-1 is observed in patients with complicated myocardial infarction [53] and stroke [54]. Furthermore, plasma ET concentrations have been reported to be strongly related to outcome after myocardial infarction [55].

Taken all together, the results that we have discussed tend to support that endothelin 1 could be indeed an atherosclerosis marker in type II diabetics and could provide prognostic information to the clinician.

In summary, decreased effective action of insulin which will manifest itself as an impaired vasodilation, together with the impaired nitric oxide production, strongly suggest that in the presence of ET-1 (although plasma ET-1 levels are normal) these could be the underlying mechanisms which promote hypertension in type II diabetes. As regards the role of ET-1 in type II hypertensive diabetics, only plasma levels have been studied [53] and no information is available on the effects of ET antagonists. Thus an exact proof of the contribution of ET-1 to the development of hypertension in diabetes is still lacking. The possible role of ET-1 in the complications in patients with diabetes mellitus needs further investigation. ET is a potential mediator of vasoconstriction and mitogenesis. As increased levels sensitize vascular segments to enhance vasoconstrictor responses we feel that our results clearly lead to the hypothesis that endothelial damage and, as suggested, the subsequent enhancement of ET-1 secretion by the diabetes per se may be in part involved in the pathogenesis of vascular complications in diabetic patients.

Hence, in summary, hypertension, insulin resistance, and NIDD have a genetic basis. The main problem that researchers face is to be able to carry out a study that will have sufficient power to discern the "genetic silhouette" through the haze of "noise" arising from the plethora of other factors that contribute to the phenotype that we call blood pressure.

The next few years should provide answers to these important questions and clarify the true functions of genotypes when exposed to excessive environmental influences.

Abstract

Diabetes mellitus is the fourth leading cause of death by disease in the United States. People with diabetes are 2–4 times more likely than nondiabetics with disease onset at middle age or later, atherosclerotic vascular disease accounts for the preponderance of deaths, irrespective of disease duration; congestive heart disease accounts for the majority of these deaths. The extent of atherosclerosis is greater in diabetics than in nondiabetics at any age.

Patients with atherosclerosis often have endothelial vasodilator dysfunction. From experimental and clinical hypertension studies it appears that alterations in endothelial function occur. This indicates that endothelial dysfunction contributes to increased peripheral vascular resistance at later stages of hypertension and most likely to the development of cardiovascular complications.

Emerging evidence suggests that endothelin-1 may be, as nitric oxide, involved in the pathogenesis of some forms of hypertension in type II diabetes mellitus. Furthermore, endothelin-1 could be indeed an atherosclerosis marker.

The results of our investigation tend to indicate that endothelin-1 levels increase with age and in type I albuminuric patients. Endothelin-1 levels were also found to be elevated in the type II diabetics either with retinopathy, albuminuria or hypertension when macrovascular disease was associated.

In summary, endothelin-1 could be an atherosclerosis marker in type II diabetics and could provide prognostic information for the clinician.

Acknowledgements. Studies of which the results are presented in part in this paper have been made possible by a grant of the Fernández-Cruz Foundation and by a further grant (SAL 90-1033) from CAICYT, Spain.

References

1. Weidmann P, Ferrari P (1991) Central role of sodium in hypertension in diabetic subjects. Diabetes Care 14:220–232
2. Ferrannini E, Buzzigoli G, Bonadonna R, Giorico MA, Oleggini M, Graziadei L, Pedrinnelli R, Brandi L, Bevilacqua S (1987) Insulin resistance in essential hypertension. N Engl Med 317: 350–357
3. Reaven GM (1988) Role of insulin resistance in human disease. Diabetes 37:1595–1607
4. Shen DC, Shieh SM, Fuh MMT, Wu D-A, Chen Y-DI, Reaven GM (1988) Resistance to insulin-stimulated glucose uptake in patients with hypertension. J Clin Endocrinol Metab 66:580–583
5. Tarn AC, Drury PL (1986) Blood pressure in children, adolescent and young adults with type 1 (insulin-dependent) diabetes. Diabetologia 29:275–281
6. Mathieson ER, Ronn B, Jensen T, Storm B, Dechkert T (1990) Relationship between blood pressure and urinary albumin excretion in development of microalbuminuria. Diabetes 39:245–249
7. Norgaar K, Feldt-Rasmussen B, Borch-Jensen K, Saelan H, Deckert T (1990) Prevalence of hypertension in insulin-dependent diabetes mellitus. Diabetologia 33:407–410
8. Viberti GC (1991) Diabetic nephropathy: links with hypertension and atherosclerosis. Cardiovasc Risk Fact 1:368–374
9. Epstein M, Sowers JR (1992) Diabetes mellitus and hypertension. Hypertension 19:403–418

10. Feldt-Rasmussen B, Mathiesen ER, Decken T, Giese J, Christensen NJ, Bent-Hansen L, Nielsen MD (1987) Central role for sodium in the pathogenesis of blood pressure changes independent of angiotensin, aldosterone and catecholamines in type-1 (insulin-dependent) diabetes mellitus. Diabetologia 30:610–617
11. Guyton AC (1991) Blood pressure control -special role of the kidneys and body fluids. Science 252:1813–1816
12. Sanchez-Martos J, Gutsowska J, Lopez-Novoa JM, Luque-Otero M, Fernandez M, Diaz Y, Ibarra J, Fernandez-Cruz A (1993) Volume expansion, natriuresis and atrial natriuretic peptide in type I and type II diabetes. Cardiovasc Risk Fact 3:174–180
13. Yuan C, Manunta P, Chen S, Hamlyn J, Pamnani M (1992) Hypertension induced by chronic ouabain administration in rats. FASEB J 6:A945
14. Manunta P, Rogowski AC, Hamilton BP, Hamlyn JM (1992) Chronic reversible hypertension in rats induced by ouabain but not digoxin. Hypertension 20:404 (abstract)
15. Manunta P, Evans G, Hamilton BP, Gann D, Resau J, Hamlyn JM (1992) A new syndrome with elevated plasma ouabain and hypertension secondary to an adrenocortical tumor. J Hypertens 10 [Suppl 4]:S27 (abstract)
16. Vague J (1956) The degree of masculine differentiation of obesities: a factor determining predisposition to diabetes, atherosclerosis, gout, and uric calculous disease. Am J Clin Nutr 4:20–34
17. Manicardi V, Camellini L, Bellodi G, Coscelli C, Ferrannin E (1986) Evidence for an association of high blood pressure and hyperinsulinemia in obese man. J Clin Endocrinol Metab 62:1302–1304
18. Modan M, Halkin H, Almog S, Lusky A, Eshkol A, Shefi M, Shitrit A, Fuchs Z (1985) Hyperinsulinemia: a link between hypertension, obesity and glucose intolerance. J Clin Invest 75:809–817
19. Storlien LH, Kraegen EW, Chisholm DJ, Ford GL, Bruce DG, Pascoe WS (1987) Fish oil prevents insulin resistance induced by high-fat feeding in rats. Science 237:885–888
20. Hall JE, Coleman TG, Mezelle HL, Smith MJ (1990) Chronic hyperinsulinemia and blood pressure regulation. Am J Physiol 258:F722–F731
21. Anderson EA, Hoffman RP, Balon TW, Sinkey CA, Mark AL (1991) Hyperinsulinemia produces both sympathetic neural activation and vasodilation in normal humans. J Clin Invest 87:2246–2252
22. Sowers JR, Khoury S, Standley P, Zemel P (1991) Mechanism of hypertension in diabetes. Am J Hypertens 4:177–182
23. Sowers Jr, Standley P, Jacoper S, Resnick M (1993) Diabetes mellitus and hypertension: the double whammy. Cardiovasc Risk Fact 3:164–169
24. Anderson E, Mark A (1993) Cardiovascular and sympathetic actions of insulin: the insulin hypothesis of hypertension revisited. Cardiovasc Risk Fact 3:159–163
25. Kramer HJ, Meyer-Lehnert H, Michel H, Predel H (1991) Endogenous natriuretic and ouabain-like factors: their roles in body fluid volume and blood pressure regulation. Am J Hypertens 4:81–89
26. Furchgott RF (1990) Studies on endothelium-dependent vasodilation and the endothelium-derived relaxing factor. Acta Physiol Scand 139:257–270
27. Kamata K, Miyata N, Kasuya Y (1989) Impairment of endothelium-dependent relaxation and changes in levels of cyclic GMP in aorta from streptozotocin-induced diabetic rats. Br J Pharmacol 97:614–618
28. Harris KH, MacLeod KM (1988) Influence of the endothelium on contractile responses of arteries from diabetic rats. Eur J Pharmacol 153:55–64
29. Cozza EN, Gomez-Sanchez CE, Foecking MF, Chiou S (1989) Endothelin binding to cultured calf adrenal zona glomerulose cells and stimulation of aldosterone secretion. J Clin Invest 84:1032–1035
30. Boarder MR, Marriott DB (1988) Characterization of endothelin-1 stimulation of catecholamine release from adrenal chromattin cells. J Cardiovasc Pharmacol 13 [Suppl 5]:S233–S224
31. Vierhapper H, Wagner O, Nowotny P, Waldhäusl W (1990) Effect of endothelin 1 in man. Circulation 81:1415–1418
32. Ohno A, Naruse M, Kato S, Hosaka M, Naruse K, Demura H (1992) Endothelin-specific antibodies decrease blood pressure and increase glomerular filtration rate and renal plasma flow in spontaneously hypertensives rats. J Hypertens 10:781–785
33. Nishikibe M, Ikada M, Tsuchida S, Fukuroda T, Shimamoto K, Kobayashy M, Ishikawa K, Yano M, Ikemoto F (1992) Antihypertensive effect of a newly synthesized endothelin antagonist, BO-123, in genetic hypertension models. J Hypertens 10 [suppl 4]:P53

34. Clozel M, Breu V, Burri K, Cassal JM, Fischli W, Gray GA, Hirth G, Loeffler BM, Mueller M, Neidhart W, Ramuz H (1993) Pathophysiological role of endothelin revealed by the first orally active endothelin receptor antagonist. Nature 365:759–761
35. Kohno M, Yasunari K, Murakawa K, Yokokawa K, Horio T, Fukui T, Takeda T (1990) Plasma immunoreactive endothelin in essential hypertension. Am J Med 88:614
36. Lüscher TF, Seo BG, Bhuler FR (1993) Controversy on endothelin in hypertension. Hypertension 21:752–757
37. Myauchi T, Yanagisawa M, Suzuki N, Iida K, Sugishita Y, Fujino M, Saito T, Goto K, Masaki T (1989) Venous plasma concentrations of endothelin in normal and hypertensive subjects (abstract). Circulation 80 [Suppl II]:II–2280
38. Davenport AP, Ashby MJ, Easton P, Ella S, Bedford J, Dickerson C, Nunez DJ, Capper SJ, Brown MJ (1990) A sensitive radioimmunoassay measuring endothelin-like immunoreactivity in human plasma: comparison of levels in patients with essential hypertension and normotensive control subjects. Clin Sci 78:261
39. Schrader J, Teebe U, Borries M, Ruschitzka F, Schoel G, Kandt M, Warneke G, Zurchner C, Weber MH, Neu U, Rath W, Henning HV (1990) Plasma Endothelin bei normalen Personen und Patienten mit nephrologisch-rheumatologischen und kardiovaskularen Erkrankungen. Klin Wochenschr 68:774
40. Schiffrin EL, Thibahult G (1991) Plasma endothelin in human essential hypertension. Am J Hypertens 4:303
41. Shihiri M, Mirata Y, Ando K, Emori T, Ohta K, Kimoto S, Ogura M, Inque A, Marumo F (1990) Plasma endothelin levels in hypertension and chronic renal failure. Hypertension 15:493
42. Saito Y, Nakao K, Mukoyama M, Shirakami G, Ito H, Yamada T, Arai H, Hosoda K, Suga S, Jougasaki M, Ogawa Y, Nakajima S, Veda M, Imura H (1990) Application of monoclonal antibodies for endothelin to hypertensive research. Hypertension 15:734
43. Noth RH, Lassman N, Tan SY, Fernandez-Cruz A, Mulrow PJ (1977) Age and the renin-aldosterone system. Arch Intern Med 137:1414–1417
44. Fernandez-Cruz A, Martin P, Fernandez L, Sanchez J, Ibarra J, Moya J, Fernandez-Durango R, Gonzalez G, Martell N, Fernandez-Pinilla C, Luque Otero M (1993) Plasma endothelin is increased in young essential hypertensives but not in elderly essential nor in diabetic hypertensives. J Hypertension 11 [Suppl 5]: S146–S147
45. Takahashi K, Ghatei M. A, Lam H.C, O'Halloran DJ, Bloom SR (1990) Elevated plasma endothelin in patients with diabetes mellitus. Diabetologia 33:306–310
46. Kawamura M, Ohgawara H, Naruse M, Suziki N, Iwasaki N (1992) Increase plasma endothelin in NIDDM patients with retinopathy. Diabetes Care 15:1396–1399
47. Collier A, Leach JP, Mclellan A, Jardine A, Morton JJ, Small M (1992) Plasma endothelinlike immunoreactivity levels in IDDM patients with microalbuminuria. Diabetes Care 8:1038–1040
48. Lee YJ, Shin SJ, Tsai JH (1994) Increased urinary endothelin-1-like immunoreactivity excretion in NIDDM patients with albuminuria. Diabetes Care 4:263–266
49. Lerman A, Edwards BS, Hallett JW, Heublein DM, Sondberg SM, Burnett JC Jr (1991) Circulating and tissue endothelin immunoreactivity in advanced atherosclerosis. N Engl J Med 321:997–1001
50. Patiño F, Ibarra J, Molino A, Fernandez-Durango R, Moya J, Fernandez-Cruz A (1994) Increased plasma endothelin in diabetes: an atherosclerosis marker? Diabetologia 37:334
51. Haak T, Jungmann E, Felber A, Hillmann U, Henning K (1992) Increased plasma levels of endothelin in diabetic patients with hypertension. Am J Hypertens 5:161–166
52. de Juan JA, Moya J, Garcia de la Cova M, Fernandez-Durango R, Fernandez-Cruz A (1993) Identification and characterization of endothelin receptor subtype (ETb) in rat retina. J Neurochem 61: 1113–1119
53. Stewart DJ, Kubac G, Costello KB, Cernacek P (1991) Increased plasma endothelin 1 in the early hours of acute myocardial infarction. J Am Coll Cardiol 18:38
54. Estrada V, Tellez MJ, Moya J, Fernandez-Durango R, Egido J, Fernandez-Cruz A (1994) High plasma levels of endothelin 1 and atrial natriuretic peptide in patients with acute ischemic stroke. Am J Hypertens 7:1085–1089
55. Omland T, Lie RT, Aakvaag A, Arsland T, Dickstein K (1994) Plasma endothelin determination as a prognostic indicator of 1-year mortality after acute myocardial infarction. Circulation 89: 1573–1579

Endothelium-Dependent Vasodilation in the Peripheral Circulation in Humans In Vivo: Effects of Hypertension and Aging

S. Taddei and A. Salvetti

Introduction

Since the pioneering work by Furchgott and Zawadzki [1], an impressive amount of evidence has clearly demonstrated in different animal specimens that relaxation of isolated arteries induced by acetylcholine requires the presence of intact endothelium. Consequently, it has been well documented that endothelial cells play a key role in the local regulation of vascular tone by production and release of vasodilator substances, including endothelium-derived relaxing factor (EDRF; identified as nitric oxide, NO), prostacyclin, and an endothelium-derived hyperpolarizing factor (EDHF). Moreover, endothelium can produce vasoconstrictor substances such as cyclooxygenase-dependent endothelium-derived contracting factors (EDCF) (prostaglandin H_2 and thromboxane A_2) and endothelin [2].

It is well documented that aging and hypertension are associated with an increased incidence of cardiovascular mortality and morbidity [3, 4]. Since most of the cardiovascular complications are related to alterations in vascular structure and function, and an impairment in endothelial function is associated with both aging [2] and hypertension [5], endothelial dysfunction can therefore be considered as one of the most important physiopathological substrates that play a role in the development of cardiovascular disease.

This review discusses the regulatory role of endothelium in human forearm circulation, focusing on dysfunctions occurring in hypertension and aging.

Endothelium and Peripheral Vascular Control in Normal Humans

To evaluate whether endothelial cells modulate vascular tone in humans, investigators used a classical approach, studying vascular response to compounds which can activate or inhibit endothelial function in the forearm or coronary circulation. To avoid reflexogenic systemic modifications caused by drug administration, agonists were directly infused into the brachial or coronary artery. Thus Vallance et al. [6] demonstrated in normotensive humans that intrabrachial infusion of N^G-monomethyl-L-arginine (L-NMMA), a competitive antagonist for NO synthase [7] (the enzyme which produces NO from the conversion of L-arginine into citrulline inside endothelial cells) reduces forearm blood flow, demonstrating that endothelium-derived NO is continuously released in human vasculature. Moreover, in the same subjects, L-NMMA blunted the vasodilating effect of acetylcholine, further confirming that even in human vessels the muscarinic agonist causes vasodilation through endothelium-dependent mechanisms. Taken together, this evidence suggests that endothelial cells can play a major role in modulating either basal blood flow or receptor-mediated vasodilator responses to agonists. These effects are at least partially mediated by NO, since, at the present time, other endothelium-dependent mechanisms involved in the control of vascular tone of normal subjects have not yet been identified. Indeed, in a group of normotensive individuals, we tried to evaluated the possibility that a cyclooxygenase-dependent EDCF might be produced in the forearm vasculature [8]. However, the failure of indomethacin, a cyclooxygenase inhibitor, to affect acetylcholine vasodilation produced convincing evidence that endothelium-derived vasoconstrictor prostanoids do not participate in the physiological modulation of vascular reactivity. Moreover, we also explored the possibility of EDHF production, and in another group of normal humans we tested whether ouabain, an inhibitor of smooth muscle cell hyperpolarization, could alter vascular response to acetylcholine [9]. The finding that the inhibitor of (Na^+/K^+)-ATPase did not affect the forearm vasodilation response to the endothelium-dependent agonist seems to exclude the possibility that, in these experimental conditions, endothelial cells can relax smooth muscle cell through the production of an EDHF.

Endothelium-Dependent Vasodilation in Human Hypertension

Dysfunction

It is well documented that in most conditions endothelium-dependent relaxations are curtailed in animal models of hypertension, possibly by a defect of synthesis and/or release of NO or by EDCF production [2, 5].

In humans, there is widespread agreement that essential hypertension is characterized by a blunted forearm vascular response to acetylcholine or methacholine, whereas the vasodilating effect of sodium nitroprusside, a direct smooth muscle cell relaxant compound, is not altered (Fig. 1) [8, 10–19]. Moreover similar results have been also demonstrated in the forearm vascular bed of patients with hypertension secondary to primary aldosteronism or renovascular hypertension. Finally, in the coronary circulation of patients with essentiall hypertension, all studies found a paradoxical vasoconstriction to acetylcholine in epicardial coronary arteries and reduced vasodilator response in coronary microcirculation, particularly in the presence of left ventricular hypertrophy [20–23]. Thus impaired endothelium-dependent vasodilation can be detected in human hypertension. However, this finding is not universal, since a study with negative results was published demonstrating no differences in forearm vasodilation responses to acetylcholine between hypertensive patients and normotensive controls [24]. This discrepancy may have its origin in the selection of the study population. Thus, in a recent study [19] we examined a large population of patients with essential hypertension and normotensive subjects, confirming that the response to acetylcholine was statistically blunted in hypertensives as compared to controls. However, only in around 10% of hypertensive patients was the curtailed vasodilating effect of the muscarinic agonist below the lower limit of the 95% con-

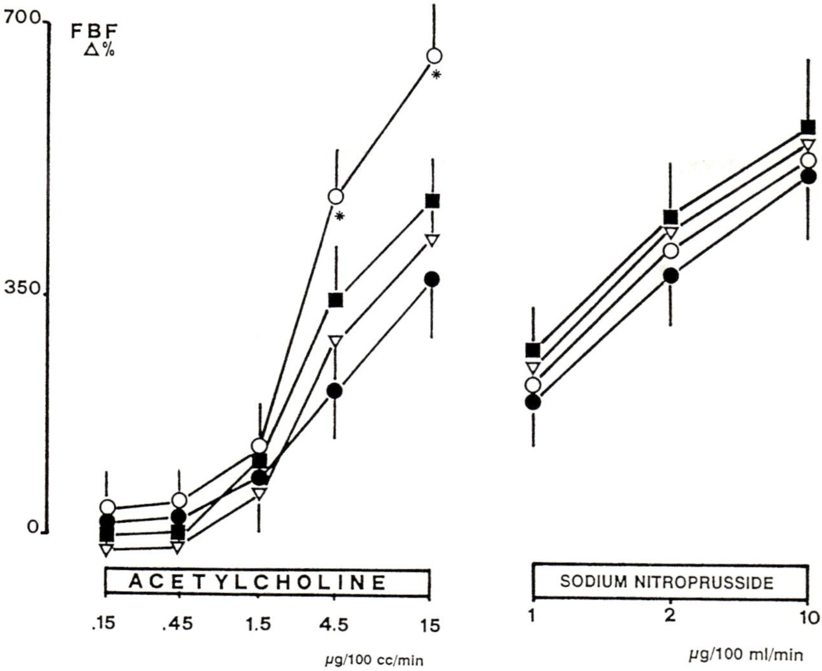

Fig. 1. Forearm vasodilation induced by intra-arterial acetylcholine (*left*) and sodium nitroprusside (*right*) in normotensive subjects (○; $n=12$), in essential hypertensive patients (●; $n=12$), in hypertensive patients with primary aldosteronism (■; $n=8$), and in renovascular hypertensive patients (▽; $n=8$). The *asterisks* show significant differences between normotensive subjects and hypertensive patients ($p<0.05$ or less). It is evident that, while response to acetylcholine is significantly blunted in hypertensive patients as compared to normotensive controls, the vasodilating effect of sodium nitroprusside was similar in the four study groups. *FBF*, forearm blood flow. (Adapted from [8], with permission)

fidence intervals of normal subject response, suggesting that several patients with essential hypertension show a response to acetylcholine that, albeit reduced, is still in a normal range. We can therefore conclude that, as shown in Table 1, an impaired vasodilator effect of acetylcholine, probably caused by endothelial dysfunction, has been extensively demonstrated in the forearm circulation of patients with essential hypertension. In addition, since this finding is not universal, comparison with a control population must always be considered in order to characterize the response.

Another tool which has been employed to test endothelial function in human hypertension is the evaluation of the effect of L-NMMA on basal forearm blood flow. Indeed Calver et al. [25] demonstrated that the vasoconstrictor response to the NO synthase inhibitor is blunted in the forearm of patients with essential hypertension as compared to controls, whereas the response to norepinephrine was not statistically different between the two groups. Therefore, in addition to agonist-evoked endothelium-dependent vasodilation, human essential hypertension is also characterized by an abnormality of endothelium-dependent basal NO-mediated vasodilation.

Table 1. Effects of intra-arterial infusion of endothelium-dependent vasodilators on forearm blood flow in patients with essential hypertension, compared to normotensive controls in studies available in literature at the present time

Author	No controls (*n*)	HT patients (*n*)	Agonist
Positive results			
Linder et al. [10]	8	8	Acetylcholine
Panza et al. [11]	18	18	Acetylcholine
Hirooka et al. [12]	14	12	Acetylcholine
Taddei et al. [8]	12	12	Acetylcholine
Panza et al. [13]	10	11	Acetylcholine
Panza et al. [14]	12	14	Acetylcholine
Panza et al. [15]	15	15	Acetylcholine
Taddei et al. [16]	13	13	Acetylcholine
Creager et al. [17]	20	24	Methacholine
Panza et al. [18]	12	14	Acetylcholine, Bradykinin
Taddei et al. [19]	53	57	Acetylcholine
Negative results			
Cockroft et al. [24]	19	17	Carbachol
	18	41	Acetylcholine

NT, normotensive; HT, essential hypertension

Mechanisms Responsible for Impaired Endothelium-Dependent Vasodilation in Human Hypertension

The evidence obtained in animal studies suggests that endothelial dysfunction can be caused either by an alteration in the L-arginine–NO pathway leading to reduced NO production and/or release and consequently to decreased endothelium-dependent relaxations [2–5] or by receptor-mediated release of vasoconstrictor products (EDCF) of cyclooxygenase, most likely prostaglandin endoperoxides such as prostaglandin H_2 [2–5].

In human essential hypertension, both alterations seem to be present. In patients with essential hypertension, but not in normotensive controls, infusion of indomethacin, a cyclooxygenase inhibitor, into the brachial artery increases forearm vasodilation to acetylcholine [8] (Fig. 2). It is worth noting that, in agreement with experimental data, indomethacin normalizes vascular response to the highest doses of the muscarinic agonist without altering the effect of the lower ones. This finding indicates that cyclooxygenase-dependent EDCF blunts the endothelium-dependent vasodilation induced by high concentrations of acetylcholine. However, different mechanisms are simultaneously involved in the impaired endothelial function which is present in essential hypertension. Indeed, Panza and colleagues [13] observed that L-NMMA failed to inhibit acetylcholine-induced vasodilation in hypertensive patients, suggesting a defect in receptor-mediated activation of the L-arginine–NO pathway in human essential hypertension.

Finally, in secondary forms of hypertension such as primary aldosteronism and renovascular hypertension, indomethacin does not alter the impaired vasodilation to

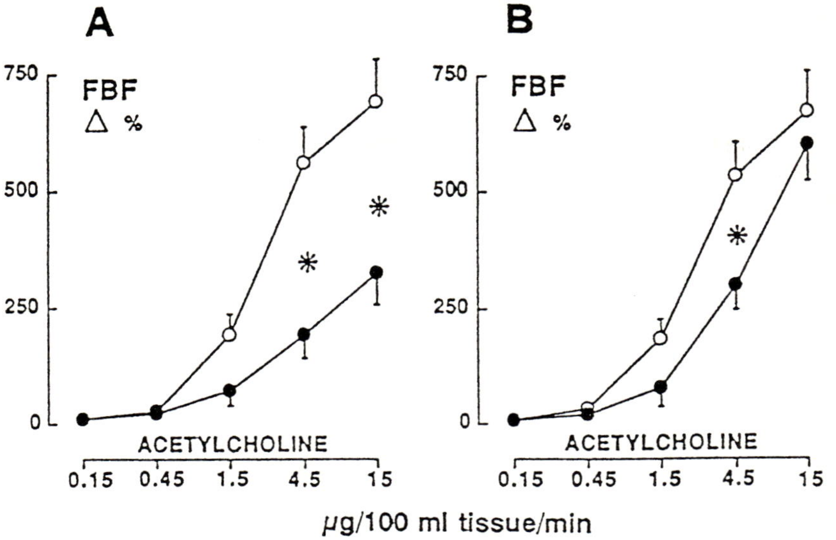

Fig. 2a, b. Forearm vasodilation induced by intra-arterial acetylcholine under control conditions (**A**, saline infusion) and in presence of intra-arterial indomethacin (**B**, 50 µg/100 ml forearm tissue/min) in normotensive subjects (○; $n=12$) and essential hypertensive patients (●; $n=12$). *Asterisks* show significant differences between normotensive subjects and hypertensive patients ($p<0.05$ or less). Cyclooxygenase inhibition by indomethacin increases vascular response to acetylcholine in patients with essential hypertension, but not in controls. *FBF*, forearm blood flow. (From [8], with permission)

acetylcholine [8], thus excluding the possibility of production of cyclooxygenase-dependent EDCF in human secondary hypertension and indicating that these constrictor prostanoids are specifically synthesized in essential hypertension. Moreover, although no available data indicate the possible mechanisms involved in impaired endothelium-dependent vasodilation in these forms of secondary hypertension, it is tempting to hypothesize a defect in the L-arginine–NO pathway.

Is Impaired Endothelial Vasodilation in Essential Hypertension a Primary or Secondary Phenomenon?

Although in animal models it appears that alterations of endothelial dysfunction in hypertension are a consequence rather than a cause of high blood pressure [26–28], findings in humans indicate different conclusions. Indeed, in the offspring of essential hypertensive parents, forearm vasodilation induced by acetylcholine is reduced compared to the offspring of normotensive individuals [29], suggesting that impaired endothelium-dependent vasodilation seems to precede the onset of essential hypertension. However, considering that only 20%–40% of subjects with a family predisposition to essential hypertension will develop the disease, the possibility that this endothelial dysfunction could contribute to the development of hypertension is only speculative. Longitudinal observations are therefore needed, especially in those relatives of hypertensives who showed a particularly blunted response (around 20% of total) to acetylcholine.

Another observation suggesting that curtailed endothelium-dependent vasodilation is not secondary to essential hypertension is that in the forearm circulation impaired vasorelaxation responses to acetylcholine do not show any correlation with blood pressure values [15, 19]. This suggests that a direct and possibly causal relationship between blood pressure and vascular endothelial responsiveness is unlikely to exist. In line with this argument is the finding that pharmacological antihypertensive treatment does not improve and/or restore endothelial function in patients with essential hypertension (see below). In contrast, in hypertensive patients with primary aldosteronism, blood pressure normalization by the surgical removal of adenoma restores endothelium-dependent responsiveness, suggesting that high blood pressure is probably the cause of the curtailed endothelial responsiveness which is associated with the disease [8].

In contrast, if we consider the basal release of NO, which, as previously discussed, is defective in patients with essential hypertension [25], the available findings seem to suggest that this endothelial dysfunction might be secondary to the development of hypertension. In fact, the blunted vasoconstrictor effect of L-NMMA shows a positive correlation with blood pressure values [25], suggesting that as blood pressure rises the NO-dependent dilator system may be less active. In line with this possibility, normalization of blood pressure values by antihypertensive treatment [30] restores a normal forearm arterial responsiveness to L-NMMA, confirming the hypothesis that the impaired basal NO release is a consequence of established hypertension.

Effect of Antihypertensive Treatment

Experiments in animal models of hypertension indicate that endothelial dysfunction can be reversed by nonpharmacological treatment, e.g., using fish oil [31] or potassium supplementation [32]. In addition, antihypertensive treatment also improves or normalizes endothelium-dependent relaxations [27, 33]. Drug effects are mainly related to changes in arterial blood pressure, since most chemically unrelated compounds exert a similar effect.

As far as nonpharmacological treatment is concerned, intrabrachial potassium chloride infusion increases vasodilation to acetylcholine, but not to sodium nitroprusside, in patients with essential hypertension [16]. This effect is abolished by simultaneous administration of L-NMMA (Fig. 3) and is not present in normotensive controls. Thus potassium seems to improve endothelium-dependent vasodilation in human essential hypertension by restoring the defective L-arginine–NO pathway. This finding could represent a likely explanation for the beneficial effects of potassium supplementation in lowering blood pressure values [34] or reducing cerebrovascular ischemic events [35].

As far as pharmacological therapy is concerned, at the present time few data are available evaluating the effect of antihypertensive treatment on endothelium-dependent vasodilation. While acute administration of the angiotensin-converting enzyme (ACE) inhibitor captopril (25 mg per os), but not of the calcium entry blocker nifedi-

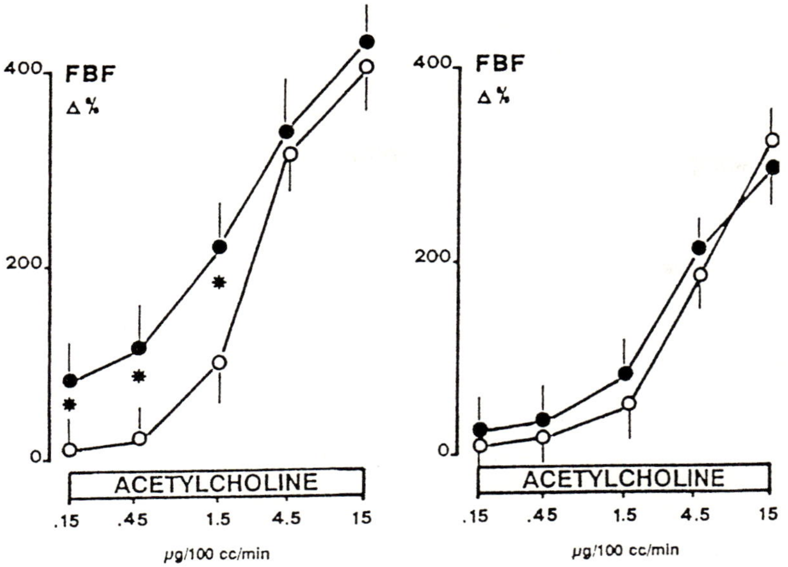

Fig. 3. Forearm vasodilation induced by intra-arterial acetylcholine in the absence (○) and presence (●) of potassium chloride (0.2 mmol/min) under control conditions (*left*) and in the presence of intra-arterial N^G-monomethyl-L-arginine (L-NMMA; 100 μg/100 ml forearm tissue per min; *right*) in essential hypertensive patients ($n=7$). Asterisks show significant differences between infusion with and without potassium ($p<0.05$ or less). Note that, in the presence of saline, potassium increases the vasodilating effect to acetylcholine and this effect is abolished by L-NMMA. *FBF*, forearm blood flow. (From [16], with permission)

pine (10 mg per os), enhanced the vasodilating response to acetylcholine in the forearm of patients with essential hypertension [12], two other studies failed to demonstrate a beneficial effect of chronic antihypertensive treatment on endothelial function. Indeed, Panza and colleagues [15] showed that blood pressure normalization caused by antihypertensive therapy with various different compounds (verapamil, β-blockers, ACE inhibitors, clonidine, and methyldopa) did not restore the impaired endothelium-dependent vascular relaxation of patients with essential hypertension. In addition, Creager et al. [17] confirmed that up to 7–8 weeks of therapy with two chemically unrelated ACE inhibitors, captopril and enalapril, did not improve the endothelium-dependent vasodilation response to acetylcholine in hypertensive humans. However preliminary results obtained in our laboratory demonstrate that the ACE inhibitor lisinopril, given either acutely (7 h after dosing) or chronically (4 weeks; 20 mg per os), enhances the response to bradykinin but not to acetylcholine, an effect probably caused by the increase in concentrations of bradykinin following inhibition of ACE, which also catalyzes the breakdown of bradykinin itself. In contrast, as already mentioned above, in patients with essential hypertension antihypertensive treatment with enalapril (10 mg day per os) or amlodipine (5 mg day per os) for 6 weeks normalizes blood pressure values and the altered vascular response to the infusion of L-NMMA [30]. Since this effect is shared by different compounds (an ACE inhibitor and a calcium entry blocker), the beneficial action of therapy on basal NO release is probably related to blood pressure normalization and not to any pharmacological interaction.

Therefore, the available information, although scanty, suggests that in human hypertension, endothelial dysfunction is much more difficult to reverse than in animal models, indicating that the defect is either primary or becomes irreversible once the hypertensive process has become stable.

Endothelium-Dependent Vasodilation and Aging in Humans

In animal vessels, controversial results have been reported on aging and endothelium-dependent relaxations. However, most studies have demonstrated an impairment of endothelial responsiveness associated with advancing age [2, 36–39].

In humans, the relationship between increasing age and endothelium-dependent vasodilation was first evaluated in coronary arteries in vivo by intra-arterial infusion of acetylcholine. All studies agreed that aging reversed the vascular response to acetylcholine in epicardial coronary arteries (left anterior descending artery or left circumflex artery) from dilation to constriction [20, 40–42]. In addition, when coronary blood flow was also measured using the Doppler catheter technique, the vasodilating effect induced by acetylcholine showed an inverse correlation with increasing age [20, 42]. Thus in humans endothelial function in coronary arteries seems to be decreased with aging. However, human coronary arteries can be affected by early atherosclerosis, even though angiographically they appear to be free from any lesions. Since atherosclerosis itself increases in old age, this possibility might potentially affect the response to acetylcholine in older patients. Therefore, we decided to evaluate whether aging can alter endothelial function in the human forearm vessels, a vascular district which is usually not affected by atherosclerosis. The results have been published in detail elsewhere [18]. In addition, to study whether advancing age can act as an independent factor in impairing endothelial function, we selected both normotensive subjects ($n=53$; age range, 19–79 years) and essential hypertensive patients ($n=57$; age range, 20–78 years). To exclude any confounding effects caused by other pathological conditions which can affect endothelium-dependent vasodilation, such as elevated plasma cholesterol, glucose levels, or smoking, hypercholesterolemic or diabetic subjects and heavy smokers were excluded from the study. In addition, both normotensives and hypertensives were selected in order to have comparable hemodynamic and humoral variables for all the age profiles. Endothelium-dependent and -independent vasodilation was assessed by the infusion of acetylcholine and sodium nitroprusside, respectively. Response to acetylcholine, but not to sodium nitroprusside, was significantly blunted in essential hypertensive patients as compared to normotensive controls. However, in both groups of subjects, the vasodilating response to acetylcholine progressively declined with increasing age, while the response to sodium nitroprusside was slightly increased only in the youngest subjects as compared to the rest of the study population. As a consequence, multivariate analysis showed a highly significant inverse correlation between age and acetylcholine, but not between age and sodium nitroprusside, in induced forearm vasodilation both in normotensive subjects (Fig. 4) and patients with essential hypertension (Fig. 5). It was noteworthy that younger normotensive and hypertensive individuals (<40 years) showed a significant and similar inverse correlation between age and vasodilation to acetylcholine (r, -0.56, $p<0.01$; r, -0.74, $p<0.001$, respectively) while older subjects (>60 years) showed no correlation (r, -0.077, not significant; r, -0.04, not significant, respectively). Therefore, these results lead us to the following conclusions: (a) endothelium-dependent vasodilation progressively decreases with advancing age; (b) the finding that this effect is also observed in patients with essential hypertension, a pathological condition which per se impairs endothelial function, indicates that aging is an independent factor which, by itself, causes progressive de-

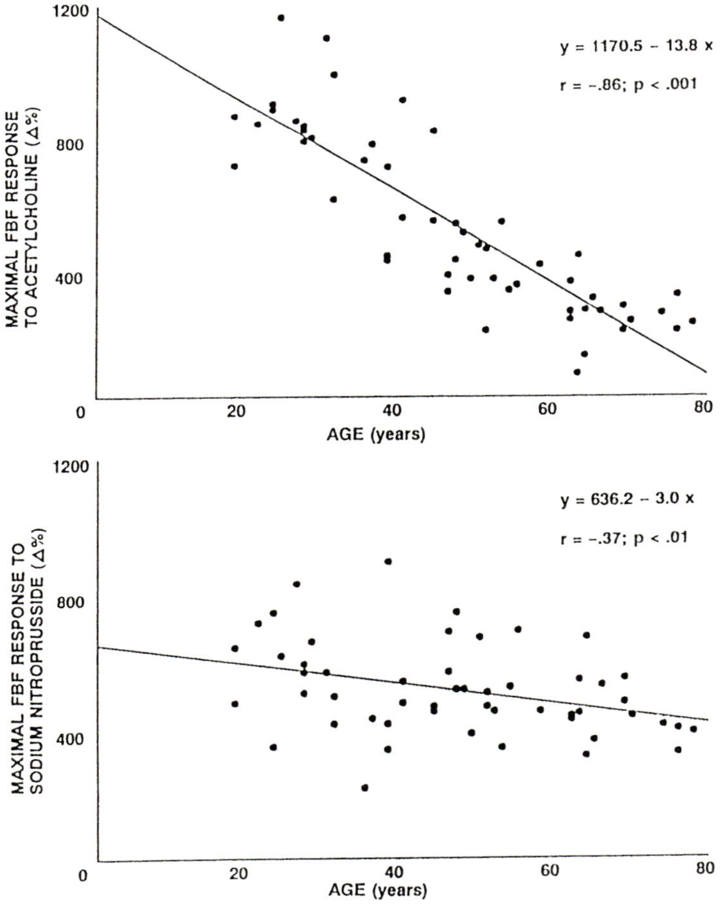

Fig. 4. Scatterplot of the relationship between age (*x-axis*) and maximum forearm blood flow response to acetylcholine (*y-axis*; *top*) and sodium nitroprusside (*y-axis*; *bottom*) in normotensive controls ($n=53$). The maximum effect is calculated as percentage increase above baseline. It is evident that a statistically significant negative correlation exists between advancing age and endothelium-dependent vasodilation responses to acetylcholine, but not between advancing age and endothelium-dependent vasodilation responses to sodium nitroprusside. (Modified from [19], with permission)

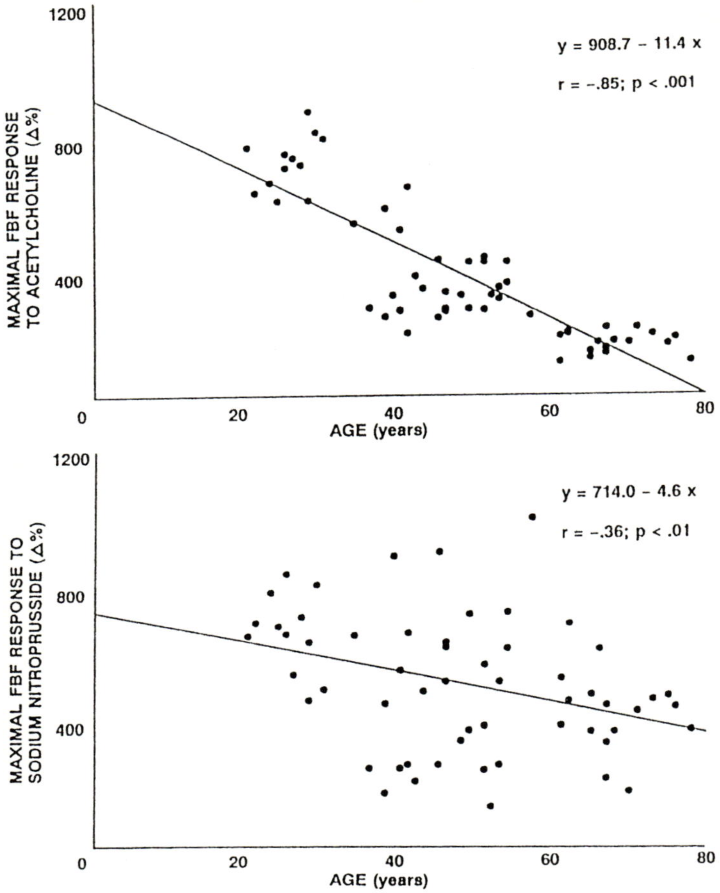

Fig. 5. Scatterplot of the relationship between age (*x-axis*) and maximum of forearm blood flow response to acetylcholine (*y-axis*; *top*) and sodium nitroprusside (*y-axis*; *bottom*) in patients with essential hypertension ($n=57$). The maximum effect is calculated as a percentage increase above baseline. It is evident that a statistically significant negative correlation exists between advancing age and endothelium-dependent vasodilation responses to acetylcholine, but not between advancing age and endothelium-independent vasodilation responses to sodium nitroprusside. (Modified from [19], with permission)

rangement of endothelium-dependent vascular control; and (c) age-related impairment in endothelial function is an early event which progressively curtails vascular function up to the age of 60 years, when the derangement seems to stop.

As far as the mechanism or mechanisms responsible for age-associated impairment of endothelial function are concerned, recent results from Küng and Lüscher [39] in aorta from normotensive Wistar-Kyoto rats (WKY) and spontaneously hypertensive rats (SHR) demonstrated that aging can determine both alterations in the release of the L-arginine–NO pathway and EDCF production. Whether these mechanisms are at work in humans has yet to be evaluated.

In conclusion, available data indicate that endothelial dysfunction is associated with advancing age in humans. This alteration could be a possible explanation of the higher cardiovascular risk associated with old age.

Conclusion

In humans, endothelial cells play a primary role in modulating vascular tone. Convincing evidence indicates that human essential hypertension and secondary hypertension can be associated with impaired endothelium-dependent vasodilation, which is probably the marker of a generalized endothelial dysfunction. In essential hypertension, such a defect could be caused by the simultaneous presence of an alteration in the L-arginine–NO pathway and the production of constrictor prostanoids. Moreover, blunted acetylcholine-induced endothelium-dependent vasodilation seems to be a primary event, since it can be detected before the onset of overt hypertension, it does not correlate with blood pressure value, and it cannot be reversed by blood pressure reduction caused by pharmacological treatment. Finally, only nonpharmacological antihypertensive treatment by potassium can partially reverse altered endothelium-dependent vasodilation.

Moreover, advancing age per se can impair endothelium-dependent vasodilation both in the coronary and forearm vascular bed both in normotensive subjects and essential hypertensive patients.

Summary

Endothelium plays a key role in modulating vascular tone through the production of vasodilator and vasoconstrictor substances. In normotensive humans, endothelium seems to affect vascular responsiveness by the production of nitric oxide. In human essential and secondary hypertension, almost all reports (only one study with negative results) indicate that the vasodilating response to the endothelium-dependent agonist acetylcholine or to the chemically related methacholine is impaired in the forearm and coronary vasculature. This abnormal response seems to be caused by the simultaneous presence of an alteration in the L-arginine–nitric oxide pathway and the production of cyclooxygenase-derived constrictor prostanoids. In contrast to observations in animals, it is possible that this impaired endothelium-dependent vasodilation could be a primary phenomenon since it can be detected in young normotensive subjects with essential hypertensive parents. In essential hypertensive patients, nonpharmacological treatment with potassium seems to enhance endothelium-dependent vasodilation whereas, although only scanty data are available, chronic pharmacological treatment does not improve the vascular response to endothelial agonists. Finally, aging can cause endothelial dysfunction both in normotensive subjects and essential hypertensive patients. In conclusion, impaired endothelium-dependent vasodilation is associated with essential and secondary hypertension and advancing age. Thus endothelial dysfunction could play an important role in the pathophysiology of hypertension and contribute to cardiovascular complications of hypertension and aging.

References

1. Furchgott RF, Zawadzki JV (1980) The obligatory role of endothelial cells in the relaxation of arterial smooth muscle by acetylcholine. Nature 288:373–376
2. Lüscher TF, Vanhoutte PM (1990) The endothelium: modulator of cardiovascular function. CRC Press, Boca Raton, pp 1–215
3. Roberts WC (1980) The hypertensive diseases. In: Laragh J (ed) Topics in hypertension, vol 17. Yorke, New York, pp 368–388
4. Kannel WB, Gordon T, Schwartz MJ (1989) Systolic versus diastolic blood pressure and risk of coronary heart disease: the Framingham study. Am J Cardiol 7:335–346
5. Vanhoutte PM (1989) Endothelium and control of vascular function. State of the art lecture. Hypertension 13:658–667
6. Vallance P, Collier J, Moncada S (1989) Effects of endothelium-derived nitric oxide on peripheral arteriolar tone in man. Lancet ii:997–1000
7. Rees DD, Palmer RMJ, Hodson HF, Moncada S (1989) A specific inhibitor of nitric oxide formation from L-arginine attenuates endothelium-dependent relaxation. Br J Pharmacol 96:418–424
8. Taddei S, Virdis A, Mattei P, Salvetti A (1993) Vasodilation to acetylcholine in primary and secondary forms of human hypertension. Hypertension 21:929–933
9. Taddei S, Virdis A, Mattei P, Natali A, Ferrannini E, Salvetti A. Effect of insulin on acetylcholine-induced vasodilation in normotensive subjects and patients with essential hypertension. Circulation, 1995 (in press)
10. Linder L, Kiowski W, Bühler, Lüscher TF (1990) Indirect evidence for the release of endothelium-derived relaxing factor in the human forearm circulation in vivo: blunted response in essential hypertension. Circulation 81:1762–1767
11. Panza JA, Quyyumi AA, Brush JE Jr, Epstein SE (1990) Abnormal endothelium dependent vascular relaxation in patients with essential hypertension. N Engl J Med 323:22–27
12. Hirooka Y, Imaizumi T, Masaki H, Ando S, Harada S, Momohara M, Takeshita A (1992) Captopril improves impaired endothelium-dependent vasodilation in hypertensive patients. Hypertension 20:175–180
13. Panza JA, Casino PR, Kilcoyne CM, Quyyumi AA (1993) Role of endothelium-derived nitric oxide in the abnormal endothelium-dependent vascular relaxation of patients with essential hypertension. Circulation 87:1468–1474
14. Panza JA, Casino PR, Badar DM, Quyyumi AA (1993) Effect of increased availability of endothelium-derived nitric oxide precursor on endothelium-dependent vascular relaxation in normal subjects and in patients with essential hypertension. Circulation 87:1475–1481
15. Panza JA, Quyyumi AA, Callahan TS, Epstein SE (1993) Effect of antihypertensive treatment on endothelium-dependent vascular relaxation in patients with essential hypertension. J Am Coll Cardiol 21:1145–1151
16. Taddei S, Mattei P, Virdis A, Sudano I, Ghiadoni L, Salvetti A (1994) Effect of potassium on vasodilation to acetylcholine in essential hypertension. Hypertension 23:485–490
17. Creager, Roddy MA (1994) Effect of captopril and enalapril on endothelial function in hypertensive patients. Hypertension 24:499–505
18. Panza JA, Garcìa CE, Kilcoyne CM, Quyyumi AA, Cannon RO (1995) Impaired endothelium-dependent vasodilation in patients with essential hypertension. Evidence that nitric oxide abnormality is not localized to a single signal transduction pathway. Circulation 91:1732– 1738
19. Taddei S, Virdis A, Mattei P, Ghiadoni L, Gennari A, Fasile Basolo C, Sudano I, Salvetti A (1995) Aging and endothelial function in normotensive subjects and essential hypertensive patients. Circulation 91:1981–1987
20. Zeiher AM, Drexter H, Saurbier B, Just H (1993) Endothelium-mediated coronary blood flow modulation in humans. Effects of age, atherosclerosis, hypercholesterolemia, and hypertension. J Clin Invest 92:652–662
21. Treasure CB, Manoukian SV, Klein JL, Vita JA, Nabel EG, Renwick GH, Selwyn AP, Alexander RW, Ganz P (1992) Epicardial coronary artery responses to acetylcholine are impaired in hypertensive patients. Circ Res 71:776–781

22. Treasure CB, Klein JL, Vita JA, Manoukian SV, Renwick GH, Selwyn AP, Ganz P, Alexander RW (1993) Hypertension and left ventricular hypertrophy are associated with impaired endothelium-mediated relaxation in human coronary resistance vessels. Circulation 7:86–93
23. Egashira K, Suzuki S, Hirooka Y, Kai H, Sugimaki M, Imaizumi T, Takeshita A (1995) Impaired endothelium-dependent vasodilation of large epicardial and resistance coronary arteries in patients with essential hypertension. Different responses to acetylcholine and substance P. Hypertension 25:201–206
24. Cockcroft JR, Chowienczyk PJ, Benjamin N, Ritter JM (1994) Preserved endothelium-dependent vasodilation in patients with essential hypertension. N Engl J Med 330:1036–1040
25. Calver A, Collier J, Moncada S, Vallance P (1992) Effect of local intra-arterial N^G-monomethyl-L-arginine in patients with hypertension: the nitric oxide dilator mechanism appears abnormal. J Hypertens 10:1025–1031
26. Lockette WE, Otsua Y, Carrettero OA (1986) Endothelium dependent relaxations in hypertension. Hypertension 8 [Suppl II]:61–66
27. Lüscher TF, Raij L, Vanhoutte PM (1987) Effect of hypertension and its reversal on endothelium-dependent relaxations in the aorta of the rat. J Hypertens 5 [Suppl 5]:153–155
28. Dohi Y, Thiel MA, Bühler FR, Lüscher TF (1990) Activation of endothelial L-arginine pathway in pressurized mesenteric resistance arteries: effect of age and hypertension. Hypertension 15:170–175
29. Taddei S, Virdis A, Mattei P, Arzilli F, Salvetti A (1992) Endothelium-dependent forearm vasodilation is reduced in normotensive subjects with familial history of hypertension. J Cardiovasc Pharmacol 20 [Suppl 12]:S193–S195
30. Lyons D, Webster J, Benjamin N (1994) The effect of antihypertensive therapy on responsiveness to local intra-arterial N^G-monomethyl-L-arginine in patients with essential hypertension. J Hypertens 12:1047–1052
31. Shimokawa H, Lam JYT, Chesebro JH, Bowie EJW, Vanhoutte PM (1987) Effects of dietary supplementation with cold-liver oil on endothelium-dependent responses in porcine coronary arteries. Circulation 76:898–905
32. Sugimoto T, Tobian L, Ganguli MC (1988) High potassium diet protects against dysfunction of endothelial cells in stroke-prone spontaneously hypertensive rats. Hypertension 11:579–585
33. Lüscher TF, Vanhoutte PM, Raij L (1987) Antihypertensive treatment normalizes decreased endothelium-dependent relaxations in rats with salt-induced hypertension. Hypertension 9:193–197
34. Skrabal F, Aubock J, Hortnagl H (1981) Low sodium/high potassium diet for prevention of hypertension: probable mechanisms of action. Lancet 2:895–900
35. Khaw KT, Barret-Connor E (1987) Dietary potassium and stroke-associated mortality. N Engl J Med 316:235–240
36. Hongo K, Nakagomi T, Kassel NF, Sasaki T, Lehmann M, Vollmer DG, Tsukahara T, Ogawa H, Torner J (1988) Effects of aging and hypertension on endothelium-dependent vascular relaxation in rat carotid artery. Stroke 19:892–897
37. Konishi M, Su C (1983) Role of endothelium in dilator responses of SHR arteries. Hypertension 5:881–886
38. Winquist RJ, Bunting PB, Baskin EP, Wallace AA (1984) Decreased endothelium-dependent relaxation in New Zeland genetic hypertensive rats. J Hypertens 2:536–541
39. Küng CF, Lüscher TF (1995) Different mechanisms of endothelial dysfunction with aging and hypertension in rat aorta. Hypertension 25:194–200
40. Vita JA, Treasure CB, Nabel EG, McLenachan JM, Fish D, Yeung AC, Vekshtein VI, Selwyn AP, Ganz P (1990) Coronary vasomotor response to acetylcholine relates to risk factors for coronary artery disease. Circulation 81:491–497
41. Yasue H, Matsuyama K, Matsuyama K, Okumura K, Morikami Y, Ogawa H (1990) Responses of angiographically normal human coronary arteries to intracoronary injection of acetylcholine by age and segment. Possible role of early coronary atherosclerosis. Circulation 81:482–490
42. Egashira K, Inou T, Hirooka Y, Kai H, Sugimachi M, Suzuki S, Kuga T, Urabe Y, Takeshita A (1993) Effects of age on endothelium-dependent vasodilation of resistance coronary artery by acetylcholine in humans. Circulation 88:77–81

Endothelin and Pulmonary Hypertension

D. J. Stewart and D. Langleben

Mechanisms of Pulmonary Hypertension

Pulmonary hypertension (PH) can be a devastating hemodynamic disorder which carries with it grave negative prognostic connotations, regardless of the etiology of increased pulmonary vascular resistance. Like systemic hypertension, PH can be the end result of complex interactions between a variety influences, some of which can be readily identified, and others remain to be elucidated. The pulmonary bed represents a unique circulation that must receive the entire cardiac output and provide the geometric and physical conditions necessary for efficient gas exchange. One crucial aspect of pulmonary vascular physiology is the ability of the lung circulation to accommodate large variations in cardiac output and present minimal impedance to blood flow at all levels of its circulation (Voelkel and Weir 1989). In patients with PH, this function is compromised, usually because of a narrowing of pulmonary arteries, either due to an increase in vascular tone (i.e. vasoconstriction) or by remodelling, with hypertrophy of medial and intimal layers and muscularization of small arterioles (Voelkel and Weir 1989).

The mechanism of PH is sometimes obvious, as in the case of PH associated with recurrent thromboembolism of large pulmonary arteries. In this condition, the mechanical blockage compromises the ability of the lung vasculature to accommodate the cardiac output. However, the precise mechanism of PH is often uncertain, even when there is a clear association with a well defined pulmonary or systemic disorder. Perhaps, a good example is PH associated with congestive heart failure. This is usually a complication of advanced heart failure, and is found in the setting of persistent and marked elevations in left ventricular filling pressures. Thus, it has been suggested that increases in transpulmonary gradient, that is the difference between the mean pulmonary artery pressure and the left atrial pressure, result from a "reactive" increase in pulmonary arteriolar tone caused by the chronic increase in capillary pressure. However, at present it is unclear how such a response might be transduced, and what specific stimuli might interact in the development of this complication. What is certain is that the appearance of PH foreshadows a poor outcome, and may contribute negatively to the overall prognosis of patients with advanced heart failure. Moreover, the presence of this complication can preclude the use of more definitive therapeutic options in this disease, namely cardiac transplantation.

Primary Pulmonary Hypertension

In some patients, the diagnosis of PH is made in the absence of any detectable cause. These patients are said to have idiopathic or primary PH and strict criteria have been established, designed to exclude known secondary causes of PH (Rich et al. 1987; Rubin 1993). However, it must be borne in mind that the etiology of PH even in this selected subgroup of patients remains obscure and it is likely that no single mechanism will be found to explain the development of pulmonary vascular abnormalities in this group. At least three major pathological "entities" have been described in patients with primary PH diagnosed clinically, which are found in varying frequencies in different populations: pulmonary microthromboembolism; plexogenic pulmonary arteriopathy; and pulmonary veno-occlusive disease (Voelkel and Weir 1989). In 1956, Paul Wood proposed a "vasoconstrictive factor" in primary PH (Wood 1958). Physiological evidence for the presence of vasoconstriction in PH has consisted of vasodilator responses to acute and chronic administration of vasodilator agents. Recent studies have has suggested a role for long-term treatment of selected patients with vasodilators (Reeves et al. 1986; Rich et al. 1992; Barst et al. 1994). However, vasoconstriction is clearly not the only mechanism contributing to increases in pulmonary vascular resistance in PH. Pathological examination, even in the early stages of primary PH (Rao et al. 1969; Heath et al. 1987), has demonstrated pronounced medial hypertrophy and proliferating "myofibroblasts". However, up until now, the search for a single factor responsible for the clinical and pathological manifestations of this disease has been unrewarding. Many potential etiologies have been proposed ranging from the influence of unknown toxins, such as that responsible for the outbreak of PH in patients receiving the appetite suppressant. Aminorex, in the late 1960s, to factors associated with female maturation and pregnancy, accounting for the increased incidence of this disease in women. Interestingly, in a rare subset of patients, PH is familial, and inherited as an autosomal dominant trait. This raises the exciting possibility that a genetic basis of this diseases may well be uncovered in the near future (Langleben 1994), however, even in the case of familial primary PH it is likely that polygenic causes will be found, as in the case of other inheritable cardiovascular diseases, i.e. hypertrophic cardiomyopathies.

Endothelial Dysfunction and Pulmonary Hypertension

A role of endothelial dysfunction in the pathogenesis of primary pulmonary hypertension has been suspected for a number of years (Mendelsohn and Loscalzo 1992). In pulmonary hypertension associated with shunts and high blood flow and repeated exposure to endotoxin, endothelial injury or dysfunction is evident (Voelkel and Weir 1989). In an animal model of pulmonary hypertension induced by monocrotaline administration (Meyrick and Reid 1979), endothelial injury was an early event, followed by endothelial proliferation and finally the development of vascular remodelling. In humans, evidence of abnormal endothelial function has often been inferred from the circulating levels of endothelium-derived substances. Increased von Willebrand factor (vWF) antigen levels have been reported in patients with primary

PH (Rabinovitch et al. 1987; Geggel et al. 1987) and abnormalities in circulating forms of vWF correlated well with increased immunostaining in lung biopsy tissue. More recently, an imbalance in vasoconstrictor and vasodilator prostanoids in the pulmonary circulation has been demonstrated in idiopathic PH, as assessed by the measurement of the stable metabolites of prostacyclin and thromboxane A2 (Christman et al. 1992). Therefore, with the discovery of the endothelin-1 by Yanagisawa et al. in 1988 it was natural to consider a potential role of this potent endothelium-derived vasoconstrictor and growth factor in PH, particularly in patients with primary disease. In the following sections, the evidence in favour of a contribution of endothelin in pulmonary vascular diseases will be briefly reviewed.

Effects of Endothelin-1 on the Pulmonary Vasculature

Although endothelin-1 (ET-1) was identified on the basis of its potent contractile actions in systemic arteries, this peptide may also exhibit dilator actions in the pulmonary circulation in some models (Eddahibi et al. 1993; Hasunuma et al. 1990; Minkes et al. 1990). With the identification of the various endothelin isoforms, it became apparent that the dilator effects were most noticeable with ET-3 (Crawley et al. 1992; Rodman et al. 1989), which was likely to release of dilatory autacoids from the endothelium, including prostacyclin and nitric oxide, particularly after bolus injection. It is now known that there are at least two major receptors for endothelins. The ET_A receptor is selective for ET-1, and mediates the contractile and growth promoting actions of this peptide on vascular smooth muscle (Bax and Saxena 1994). In contrast, a nonselective receptor, ET_B, is found on the endothelium, and induces the release of nitric oxide and prostacyclin. However, increasingly it has become recognized ET_B can also be expressed by pulmonary vascular smooth muscle, and can mediate direct contractile effects (Warner 1994). The development of PH in rats exposed to chronic hypoxia was associated with a loss of vasodilator actions in the isolated lungs studied in vitro (Eddahibi et al. 1993), and a potentiation of its vasoconstrictor effects. In vitro, ET-1 is a potent vasoconstrictor of isolated pulmonary arteries and veins (for recent review see Barnes 1994). In general, the potency of ET-1 in the pulmonary circulation is similar to that in systemic beds, and it appears to be particularly potent in the small, resistance arteries. Important developmental changes have been described in the neonatal period, with the constrictor of ET-1 actions becoming more prominent with increasing age (Toga et al. 1992; Liben et al. 1993; Wong et al. 1994; Krzeski et al. 1991).

Effects of Pulmonary Vascular Cell Growth and Matrix Accumulation

In addtion to its vasomotor effects, ET-1 has been demonstrated to induce proliferation of systemic smooth muscle cells and fibroblasts (Battistini et al. 1993). Like vasoconstriction, this action is mediated primarily via ET-1 selective, ET_A receptors. Although the potency of ET-1 as a growth factor is variable in different models, it does appear to significantly potentiate the effects of other growth factors (Janakidevi et al. 1992). As well, ET-1 has been reported to increase the synthesis of collagen by fibroblasts (Vancheeswaran et al. 1994; Falanga et al. 1992) and increase matrix production. Both these processes could contribute to the well known morphological features associated with PH. In smooth cells of pulmonary vascular origin, ET-1 has also been shown to increase mitosis (Janakidevi et al. 1992; Hassoun et al. 1992), indicating that this mechanism may be relevant to the pathogenesis of PH.

Pulmonary Clearance of ET-1

In addition to its crucial role in gas exchange, the lung has important metabolic functions, including the clearance of a wide variety of substances from the blood. Earlier studies on mechanisms of clearance of circulating ET-1 provided conflicting results. Following administration large doses of ET-1, resulting in supraphysiological plasma levels, the primary site of clearance appeared to be the kidney (Pernow et al. 1989). However, the importance of renal clearances at more physiological ET-1 concentrations, or even those observed in various pathophysiological states, is less certain. At these lower levels, pulmonary clearance dominates (Sirvio et al. 1990; Rimar and Gillis 1992), with negligible clearance by other beds. In the anaesthetized dog, Dupuis et al. (1994) showed a first-pass pulmonary clearance of 35% and increasing with longer transit times through the pulmonary circuit. A similar magnitude of pulmonary clearance has been observed in awake human subjects (J. Dupuis, personal communications). Recently, it has been shown that pulmonary clearance of ET-1 is mediated by ET_B (Fukuroda et al. 1994), and can be blocked by antagonist of this receptors. Thus, under physiological conditions the normal lung may act to maintain low plasma levels of ET-1 levels in a variety of cardiovascular diseases.

The Endothelin System in Pulmonary Hypertension

Soon after the discovery of ET-1, increased plasma levels were described in a number of cardiovascular disorders, pointing to a possible role in disease states characterized by increased vascular tone. The most dramatic elevations were found in extreme states, such as cardiogenic shock (Cernacek and Stewart 1989), however, highly significant increases in ET-1 levels were also found in patients with heart failure, renal failure and pulmonary hypertension. In the following sections, the mechanism of elevation of circulating ET-1 in pulmonary hypertension (i.e. increased production vs. decreased clearance) will be explored, as well as the possible relevance the increased ET-1 levels in the pathogenesis of this disease.

Experimental Models of Pulmonary Hypertension

Hypoxia has long been recognized as a stimulus for pulmonary vasoconstriction. The effect of low pO_2 on ET-1 release by cultured endothelial cells has been examined by several groups with conflicting results. Using systemic human endothelial cells, an increase in ET-1 release and prepro-ET-1 mRNA expression was found on lowering

pO_2 to 30 Torr (Kourembanas et al. 1991). However, several groups have observed no change in ET-1 production or expression in bovine pulmonary artery endothelial cells in response to hypoxia (Langleben et al. 1989; Wiebke et al. 1992; Hassoun et al. 1992), raising the possibility that the pulmonary endothelium, which is normally acclimatized to lower oxygen levels may behave differently from systemic endothelium. In vivo, chronic exposure of rats to hypoxic conditions resulting in development of PH, has been reported to increase plasma levels of ET-1 and prepro-ET-1 mRNA expression in pulmonary tissue (Elton et al. 1992; Li et al. 1994), but not in systemic vessels. As well, increased expression of ET_A and ET_B receptors were observed in the lung (Elton et al. 1992; Li et al. 1994), suggesting an exaggerated action of endogenously produced ET-1. Thus, these studies support a role for locally produced ET-1 in PH induced by chronic hypoxia, but it is unlikely that this mediator can account for the rapid and brief pulmonary vasoconstriction resulting from acute hypoxic exposure. Evidence of increased pulmonary ET-1 production has also been found in other models of PH. Stelzner et al. (1992) reported increases in plasma ET-1 and pulmonary prepro-ET-1 mRNA expression, preceding the onset of PH in a rat model of idiopathic PH. As well, greater expression of ET_B receptors was observed in the monocrotoline rat model (Yorikane et al. 1993), although this model was associated with decreased overall expression of ET-1 mRNA (Miyauchi et al. 1993). This finding is not necessarily inconsistent with a role for ET-1 in this model (see below), since monocrotaline-induced damage to pulmonary epithelium, which contains high amounts of ET-1 in the rat, might reduce total tissue levels independent of any changes in vascular expression.

Human Studies

Although animal models provide useful information about the potential role of ET in PH, they do not necessarily reproduce the disease states which are clinically relevant. Furthermore, the mechanism of PH in different diseases likely varies considerably, and the relative importance of the vasoconstrictor peptide may also differ widely. In a preliminary report, ET-1 levels were elevated in a small series of patients with primary and secondary PH (Cernacek and Stewart 1989), however, the mechanism of elevation and relevance of this finding remained to be elucidated. We attempted to address these questions by measuring the ET-1 gradient across the pulmonary bed in patients with primary or secondary PH, as well as in normal subjects and patients with stable coronary disease and normal pulmonary artery pressures (Stewart et al. 1991). There was a significant increase in mean ET-1 plasma levels in both groups of PH patients compared to control subjects, but no significant difference between primary and secondary forms of PH. However, when simultaneous systemic arterial (i.e. pulmonary effluent) and systemic venous (pulmonary inflow) levels were compared for each patient, a distinct pattern began to emerge. In the control groups, the levels in the systemic arterial blood were consistently lower than in the inflow blood, suggesting a net clearance by the healthy lung, and consistent with the experimental data reviewed above (Sirvio et al. 1990; Rimar and Gillis 1992). In patients with secondary PH, ET-1 levels were similar in the arterial and venous blood, possibly indicating decreased ET-1 clearance or a combination of increased production and decreased clearance. In contrast, in the primary PH group the levels were greater in the systemic arterial blood compared with those in the mixed venous blood in nearly all subjects. This highly significant difference strongly suggested a net production of ET-1 across the lung in this highly selected group of patients.

Other data supporting an association between ET-1 and PH in a variety of clinical situations have since been reported. Cody et al. (1992) observed that the increased ET-1 levels in patients with severe heart failure, correlating with inappropriate elevations in pulmonary vascular resistance, as assessed by the ratio of pulmonary to systemic vascular resistance. In the same condition, a relationship between the spillover of ET-1 in the pulmonary bed, calculated from the difference between the levels in blood aspirated from the pulmonary capillary wedge position and from the main pulmonary artery, and pulmonary vascular resistance was described (Tsutamoto et al. 1994), suggesting that increased local pulmonary ET-1 production may contribute to PH in patients with heart failure. As well, increased ET-1 levels were found in infants with PH associated with congenital heart defects (Yoshibayashi et al. 1991), with a progressive increment in plasma levels from right ventricle to pulmonary vein, again supporting increased pulmonary production.

Endothelial damage is also a hallmark of the adult respiratory distress syndrome (ARDS), and PH is a frequent complication of this disorder. Recent studies in ARDS have revealed elevated circulating levels of ET-1 (Druml et al. 1993; Langleben et al. 1993). The damaged lung is unable to adequately clear ET-1 from the circulation, and net ET-1 release can be seen is some patients (Druml et al. 1993; Langleben et al. 1993). Evidence of endothelial dysfunction appears within 24 h of lung injury, and improvement in the balance between ET-1 production and clearance parallels the clinically recovery from ARDS (Langleben et al. 1993). In contrast, patients whose condition worsened exhibited persistent elevations in ET-1 and abnormal pulmonary ET-1 handling (Langleben et al. 1993).

Local Pulmonary ET-1 Production in Pulmonary Hypertension

Although indirect evidence pointing to increased production of ET-1 by the pulmonary vasculature in PH has been presented, direct proof requires the detection of increased product and message in the lung. We examined the abundance of immunoreactive ET-1 in lung tissue from patients with primary plexogenic PH and with various secondary causes of PH (Giaid et al. 1993c), compared with normal lungs harvested but not used for lung transplantation or lung tissue from patients with pulmonary disease but no PH. Little or no immunoreactive ET-1 was present in the control lung tissue, whereas in PH, abundant immunostaining was detected at all levels of the pulmonary vasculature, including elastic and muscular arteries, capillaries and veins and localized primarily to the pulmonary endothelium. The greatest staining was found in the muscular pulmonary arteries, which also exhibited the greatest morphological abnormalities, specifically marked medial and intimal hyperplasia and fibrosis. Interestingly, similar levels of ET-1-like immunoreactivity were observed in the two patient groups except at the level of muscular arteries, in which there was significantly greater staining in patients with plexogenic PH. As well, only in this latter group was there a significant correlation between immunostaining grade and pulmonary vascular resistance. Increased local production of ET-1 in patients with PH was confirmed by the demonstration of increased prepro-ET-1 mRNA expression using in situ hybridization in both patient groups.

In addition to increased ET-1 expression by pulmonary vessels, a high degree of immunostaining was also observed in the systemic bronchial arteries of patients with PH. The relevance of this finding is unclear, however it might be related to the formation of collaterals given the possible role of ET-1 in angiogenesis. Indeed, in a unilateral pulmonary artery occlusion model of pulmonary vasculopathy, similar increases in ET-1-like immunoreactivity were observed associated with collateral formation (Giaid et al. 1993b). Occasionally, marked ET-1 expression was observed over nonendothelial cells, including smooth muscle cells at sites of pulmonary atherosclerosis, and hyperplastic type II alveolar pneumocytes in patients with pulmonary fibrosis with or without PH (Giaid et al. 1993a). These findings strongly support a role for ET-1 in pulmonary diseases associated with vascular and parenchymal cell proliferation and matrix production, and complement the earlier data based on plasma measurements.

ET Antagonists in Pulmonary Hypertension

The development of potent and specific inhibitors of ET receptors over the course of the last several years may open the door to an exciting new era of pharmacological therapy of cardiovascular disorders. In addition to their therapeutic potential, these agents offer an opportunity to directly test the hypothesis that endothelins are important in the pathogenesis of PH. BQ-123, a selective peptide inhibitor of ET_A significantly reduced pulmonary artery pressure and right ventricular hypertrophy following monocrotaline administration in the rat (Miyauchi et al. 1993), supporting a role for ET-1 in this model. More recently, the orally active mixed ET-receptor antagonist, bosentan, was similarly effective in protecting against increased pulmonary artery pressures and right ventricular remodelling in rats exposed to chronic hypoxia (Eddahibi et al. 1995). This study also demonstrated decreased muscularization of distal pulmonary vessels in animals treated with the ET antagonist, suggesting that in addition to reversing the vasomotor actions of ET-1, there was an inhibitory effect of smooth muscle migration and proliferation. However, no study has yet been reported using ET receptor antagonists in patients with PH. Ultimately, the results of such a trial will be necessary to assess the potential clinical utility of these agents in the treatment of this disorder.

Conclusion

It is too early to say whether the search for the "vasoconstrictor factor" in PH will end the discovery of the endothelin system. However, the evidence implicating ET-1 in the characteristic functional and morphological vascular abnormalities of PH is mounting. It is reasonable, on the basis of the data reviewed here, to conclude that there may be a special role for this potent vasoconstrictor and growth promoting peptide in vascular diseases of the lung, as well as in parenchymal pulmonary disorders characterized by abnormal cellular proliferation and matrix accumulation. The initial experience with selective endothelin antagonists in animal models of PH would support such a view, and the case for the use of these agents in patients suffering from PH, either as a primary disease or in association with other cardiovascular disorders, is becoming progressively more compelling.

Abstract

Pulmonary hypertension, both of primary and secondary causes, remains a serious and progressive clinical disorder which carries a poor prognosis. At present, the pharmacological approaches to the treatment of pulmonary hypertension are limited. The purpose of this chapter is to review the evidence that endothelin (ET-1), a potent vasoconstrictor and mitogenic peptide, contributes to the pathophysiology of pulmonary hypertension. ET-1 is a potent constrictor of pulmonary arteries and veins in vitro. In vivo, ET-1 can cause pulmonary vasodilation by interaction with ET_B receptors on endothelium, releasing vasodilator autocoids or vasoconstriction by direct interaction with ET_A as well as ET_B receptors on smooth muscle cells. Vasoconstrictor responses become more important with increase age and are augmented in animal models of pulmonary hypertension. Levels of circulating ET-1 are increased in both animal models of hypertension as well as in patients with pulmonary vascular disease. Levels of ET-1 are increased in both primary and secondary pulmonary hypertension possibly due to decreased pulmonary clearance of ET-1 as well as increase local pulmonary expression of ET-1 has been confirmed by immunostaining and in situ hybridization in animal models and in patients with pulmonary hypertension. The most consistent relationship between increased pulmonary production of ET-1 and functional vascular abnormalities is seen in patients with primary pulmonary hypertension. The initial experience using endothelin antagonists strongly supports a role for this system in pulmonary hypertension and clinical trials are urgently needed to explore the therapeutic efficacy of this exciting new treatment modality.

References

Barnes PJ (1994) Endothelins and pulmonary diseases. J Appl Physiol 77:1051–1059

Barst RJ, Rubin IJ, McGoon MD, Caldwell EJ, Long WA, Levy PS (1994) Survival in primary pulmonary hypertension with long-term continuous intravenous prostacyclin. Ann Intern Med 121: 409–415

Battistini B, Chailler P, D'Orléans-Juste P, Brière N, Sirois P (1993) Growth regulatory properties of endothelins. Peptides 14:385–399

Bax WA, Saxena PR (1994) The current endothelin receptor classification: time for reconsideration. Trends Pharmacol Sci 15:379–386

Cernacek P, Stewart DJ (1989) Immunoreactive endothelin in human plasma: marked elevations in patients in cardiogenic shock. Biochem Biophys Res Commun 161:562–567

Christman BW, McPherson CD, Newman JH, King GA, Bernard GR, Groves BM, Loyd JE (1992) An imbalance between excretion of thromboxane and prostacyclin metabolites in pulmonary hypertension. N Engl J Med 327:70–75
Cody RJ, Haas GJ, Binkley PF, Capers Q, Kelley R (1992) Plasma endothelin correlates with the extent of pulmonary hypertension in patients with chronic congestive heart failure. Circulation 85:504–509
Cawley DE, Liu SF, Barnes PJ, Evans TW (1992) Endothelin-3 is a potent pulmonary vasodilator in the rat. J Appl Physiol 72:1425–1431
Druml W, Stelzer H, Waldhäusl W, Lenz K, Hammerle A, Vierhapper H (1993) Endothelin-1 in adult respiratory distress syndrome. Am Rev Respir Dis 148:1169–1173
Dupuis J, Goresky CA, Stewart DJ (1994) Pulmonary removal and production of endothelin in the anesthetized dog. J Appl Physiol 76:694–700
Eddahibi S, Springall D, Mannan M, Carville C, Chabrier P-E, Levame M, Raffestin B, Polak J, Adnot S (1993) Dilator effect of endothelins in pulmonary circulation: changes associated with chronic hypoxia. am J Physiol Lung Cell Mol Physiol 265:L571–L580
Eddahibi S, Raffestin B, Clozel M, Levame M, Adnot S (1995) Protection from pulmonary hypertension with an orally active endothelin receptor antagonist in hypoxic rats. Am J Physiol Heart Circ Physiol 268:H828–H835
Elton TS, Oparil S, Taylor GR, Hicks PH, Yang RH, Jin H, Chen YF (1992) Normobaric hypoxia stimulates endothelin-1 gene expression in the rat. Am J Physiol Regul Integr Comp Physiol 263: R1260–R1264
Falanga V, Katz MH, Kirsner R, Alvarez AF (1992) The effects of endothelin-1 on human dermal fibroblast growth and synthetic activity. J Surg Res 52:515–519
Fukuroda T, Fujikawa T, Ozaki S, Ishikawa K, Yano M, Nishikibe M (1994) Clearance of circulating endothelin-1 by ETB receptors in rats. Biochem Biophys Res Commun 199:1461–1465
Geggel RL, Cavalho ACA, Hoyer LW, Reid LM (1987) Von Willebrand factor abnormalities in primary pulmonary hypertension. Am Rev Respir Dis 135:294–299
Giaid A, Michel RP, Stewart DJ, Sheppard M, Corrin B, Hamid Q (1993a) Expression of endothelin-1 in lungs of patients with cryptogenic fibrosing alveolitis. Lancet 341:1550–1554
Giaid A, Stewart DJ, Michel RP (1993b) Endothelin-1-like immunoreactivity in postobstructive pulmonary vasculopathy. J Vasc Res 30:333–338
Giaid A, Yanagisawa M, Langleben D, Michel R, Levy R, Shennib H, Kimura S, Masaki T, Duguid WP, Stewart DJ (1993c) Expression of endothelin-1 in lungs of patients with pulmonary hypertension. N Engl J Med 328:1732–1739
Hassoun PM, Thappa V, Landman MJ, Fanburg BL (1992) Endothelin 1: mitogenic activity on pulmonary artery smooth muscle cells and release from hypoxic endothelial cells. Proc Soc Exp Biol Med 199:165–170
Hasunuma K, Rodman DM, O'Brien RF, McMurtry IF (1990) Endothelin 1 causes pulmonary vasodilation in rats. Am J Physiol Heart Circ Physiol 259:H48–H54
Heath D, Smith P, Gosney J, Mulkahy D, Fox K, Yacoub M, Harris P (1987) the pathology of early and late stages of primary pulmonary hypertension. Br Heart J 58:204–213
Janakidevi K, Fisher MA, Del Vecchio PJ, Tiruppathi C, Figge J, Malik AB (1992) Endothelin-1 stimulates DNA synthesis and proliferation of pulmonary artery smooth muscle cells. Am J Physiol Cell Physiol 263:C1295–C1301
Kourembanas S, Maraden PA, McQuillan LP, Faller DV (1991) Hypoxia induces endothelin gene expression and secretion in cultured human endothelium. J Clin Invest 88:1054–1057
Krzeski R, Long W, Katayama H, Henry W (1991) Hemodynamic effects of endothelin-1 in the newborn piglet: influence on pulmonary and systemic vascular resistance. J Cardiovasc Pharmacol 17 [Suppl 7]:S322–S325
Langleben D (1994) Familial primary pulmonary hypertension. Chest 105:13S–16S
Langleben D, Cernacek P, Stewart DJ (1989) Moderate hypoxia and hyperoxia do not alter endothelin release from bovine pulmonary endothelium in vitro. Circulation 80:II–288 (Abstract)
Langleben D, DeMarchie M, Laporta d, Spanier AH, Schlesinger RD, Stewart DJ (1993) Endothelin-1 in acute lung injury and the adult respiratory distress syndrome. Am Rev Respir Dis 148:1646–1650
Li H, Chen S-J, Chen Y-F, Meng QC, Durand J, Oparil S, Elton TS (1994) Enhanced endothelin-1 and endothelin receptor gene expression in chronic hypoxia. J Appl Physiol 77:1451–1459

Liben S, Stewart DJ, De Marte J, Perreault T (1993) Ontogeny of Big endothelin-1 effects in newborn piglet pulmonary vasculature. Am J Physiol Heart Circ Physiol 265: H139–H145

Mendelsohn ME, Loscalzo J (1992) Endotheliopathies. In: Loscalzo J, Creager MA, Dzau VJ (eds) Vascular medicine: a textbook of vascular biology and diseases. Little, Brown, and Co., Boston, pp 279–305

Meyrick B, Reid L (1979) Development of pulmonary arterial changes in rats fed crotalaria spectabilis. Am J Path 94: 37

Minkes, RK, Bellan JA, Saroyan RM, Kerstein MD, Coy DH, Murphy WA, Nossaman BD, McNamara DB, Kadowitz PJ (1990) Analysis of cardiovascular and pulmonary responses to endothelin-1 and endothelin-3 in the anesthetized cat. J Pharmacol Exp Ther 253: 1118–1125

Miyauchi T, Yorikane R, Sakai S, Sakurai T, Okada M, Nishikibe M, Yano M, Yamaguchi I, Sugishita Y, Goto K (1993) Contribution of endogenous endothelin-1 to the progression of cardiopulmonary alterations in rats with monocrotaline-induced pulmonary hypertension. Circ Res 73: 887–897

Pernow J, Hensén A, Lundberg JM (1989) Tissue specific distribution, clearance and vascular effects of endothelin in the pig. Biochem Biophys Res Commun 161: 647–653

Rabinovitch M, Andrew M, Thom H, Trusler GA, Williams WG, Rowe RD, Olley PM (1987) Abnormal endothelial factor VIII associated with pulmonary hypertension and congenital heart defects. Circulation 766: 1043–1052

Rao BNS, Moller JH, Edwards JE (1969) Primary pulmonary hypertension in a child. Circulation 40: 583–587

Reeves JT, Grove BM, Turikevich D (1986) The case for treatment of selected patients with primary pulmonary hypertension. Am Rev Respir Dis 134: 342–346

Rich S, Dantzker DR, Ayres SM, Bergofski EH, Brundage BH, Detre KM, Fishman AP (1987) Primary pulmonary hypertension. Ann Intern Med 107: 216–223

Rich S, Kaufman E, Levy PS (1992) The effect of high doses of calcium channel blockers on survival in primary pulmonary hypertension. New Engl J Med 327: 76–81

Rimar S, Gillis CN (1992) Differential uptake of endothelin-1 by the coronary and pulmonary circulations. J Appl Physiol 73: 557–562

Rodman DM, McMurtry IF, Peach JL, O'Brien RF (1989) Comparative pharmacology of rat and porcine endothelin in rat aorta and pulmonary artery. Eur J Pharmacol 165: 297–300

Rubin LJ (1993) Primary pulmonary hypertension. Chest 104: 236–250

Sirvio M-L, Metsarinne K, Satjonmas O, Fyhrquist F (1990) Tissue distribution and half-life of 125I-endothelin in the rat: importance of pulmonary clearance. Biochem Biophys Res Commun 167: 1191–1195

Stelzner TJ, O'Brien RF, Yanagisawa M, Sakurai T, Sato K, Webb S, Zamora M, McMurtry IF, Fisher JH (1992) Increased lung endothelin-1 production in rats with idiopathic pulmonary hypertension. Am J Physiol Lung Cell Mol Physiol 262: L614–L620

Stewart DJ, Levy RD, Cernacek P, Langleben D (1991) Increased plasma endothelin-1 in pulmonary hypertension: marker or mediator of disease? Ann Intern Med 114: 464–469

Toga H, Ibe BO, Raj JU (1992) In vitro responses of ovine intrapulmonary arteries and veins to endothelin-1. Am J Physiol Lung Cell Mol Physiol 263: L15–L21

Tsutamoto T, Wada A, Maeda Y, Adachi T, Kinoshita M (1994) Relation between endothelin-1 spillover in the lungs and pulmonary vascular resistance in patients with chronic heart failure. J Am Coll Cardiol 23: 1427–1433

Vancheeswaran R, Magoulas T, Efrat G, Wheeler-Jones C, Olsen I, Penny R, Black CM (1994) Circulating endothelin-1 levels in systemic sclerosis subsets – A marker of fibrosis or vascular dysfunction. J Rheumatol 27: 1838–1844

Voelkel NF, Weir EK (1989) Etiologic mechanisms in primary pulmonary hypertension. In: Weir EK, Reaves JT (eds) Pulmonary vascular physiology and pathophysiology. Marcel Dekker, New York, pp 513–539

Warner TD (1994) Endothelin receptor antagonists. Cardiovasc Drug Rev 12: 105–122

Wiebke JL, Montrose-Rafizadeh C, Zeitlin PL, Guggino WB (1992) Effect of hypoxia on endothelin-1 production by pulmonary vascular endothelial cells. Biochim Biophys Acta Mol Cell Res 1134: 105–111

Wong J, Vanderford PA, Fineman JR, Soifer SJ (1994) Developmental effects of endothelin-1 on the pulmonary circulation in sheep. Pediatr Res 36: 394–401

Wood P (1958) Pulmonary hypertension with special reference to the vasoconstrictive factor. Br Heart J 20:557–570

Yanagisawa M, Kurihara H, Kimura S, Tomobe Y, Kobayashi M, Mitsui Y, Yazaki Y, Goto K, Masaki T (1988) A novel potent vasoconstrictor peptide produce by vascular endothelial cells. Nature 332:411–415

Yorikane R, Miyauchi T, Sakai S, Sakurai T, Yamaguchi I, Sugishita Y, Goto K (1993) Altered expression of ET_B-receptor mRNA in the lung of rats with pulmonary hypertension. J Cardiovasc Pharmacol 22 [Suppl 8]: S336–S338

Yoshibayashi M, Nishioka K, Nakao K, Saito Y, Matsumura M, Ueda T, Temma S, Shirakami G, Imura H, Mikawa H (1991) Plasma endothelin concentrations in patients with pulmonary hypertension associated with congenital heart defects: Evidence for increased production of endothelin in pulmonary circulation. Circulation 84:2280–2285

Subject Index